Germs, Genes, & Civilization

Germs, Genes, & Civilization

*How Epidemics Shaped
Who We Are Today*

David P. Clark
Department of Microbiology,
Southern Illinois University

Vice President, Publisher: Tim Moore
Associate Publisher and Director of Marketing: Amy Neidlinger
Editorial Assistant: Myesha Graham
Development Editor: Kirk Jensen
Operations Manager: Gina Kanouse
Senior Marketing Manager: Julie Phifer
Publicity Manager: Laura Czaja
Assistant Marketing Manager: Megan Colvin
Cover Designer: Sandra Schroeder
Managing Editor: Kristy Hart
Senior Project Editor: Lori Lyons
Copy Editor: Krista Hansing Editorial Services
Proofreader: Kay Hoskin
Senior Indexer: Cheryl Lenser
Compositor: Nonie Ratcliff
Manufacturing Buyer: Dan Uhrig

FT Press offers excellent discounts on this book when ordered in quantity for bulk
purchases or special sales. For more information, please contact U.S. Corporate and
Government Sales, 1-800-382-3419, corpsales@pearsontechgroup.com. For sales outside
the U.S., please contact International Sales at international@pearson.com.

Printed in the United States of America

First Printing May 2010

ISBN-10: 0-13-701996-3
ISBN-13: 978-0-13-701996-0

Pearson Education LTD.
Pearson Education Australia PTY, Limited.
Pearson Education Singapore, Pte. Ltd.
Pearson Education North Asia, Ltd.
Pearson Education Canada, Ltd.
Pearson Educación de Mexico, S.A. de C.V.
Pearson Education—Japan
Pearson Education Malaysia, Pte. Ltd.

Library of Congress Cataloging-in-Publication Data

Clark, David P.
  Germs, genes & civilization : how epidemics shaped who we are today / David P. Clark.
     p. cm.
  Includes index.
  ISBN 978-0-13-701996-0 (hardback : alk. paper) 1. Diseases and history. 2. Epidemics.
3. Civilization. 4. Human evolution. 5. Human genetics. I. Title. II. Title: Germs, genes,
and civilization.
  R702.C63 2010
  614.4—dc22

                                                                          2009048726

This book is dedicated to my younger brother, Andrew,
who has always enjoyed a good argument.

Contents

Preface

Humans typically labor under the illusion that they control their own destiny. This book argues that, in reality, invisible microbes often control human activities. Recent findings have shown that animals that develop without their natural bacterial inhabitants have defective immune systems and poor health. Thus, we and other animals depend on the bacteria for our healthy development. On a larger scale, we now recognize that microbes maintain the global ecosystem and are partly responsible for keeping our planet healthy. The amount of "good" bacteria that work to recycle nutrients and degrade waste is greater by far than the amount of "bad" bacteria that threaten human health.

Here I enter the intermediate zone between individual development and planetary ecology, to discuss how microbes have decided major historical events and shaped cultural trends. Furthermore, the emergence of resistance to infectious diseases has selected alterations in genes that affect human behavior.

This book is *not* a history of public health, medicine, or microbiology, although it does mention these issues. Instead, this book describes how infections have shaped both individual humans and their societies from the very beginning of civilization. Disease has influenced our cultural and religious beliefs, as well as determined the outcome of wars and major historical events. I have tried to show how beneficial long-term effects have resulted from epidemics that were terrible tragedies to those caught up in them.

Philosophically, we are just emerging from a period of transition between the perfectionist and selfish views of nature. The classic example, in the area of disease, is the idea that, over the ages, infectious agents will adapt to their hosts. Eventually, all diseases will become no worse than a bad cold. This is an attempt to retain a utopian future while allowing evolution to occur. Recently, we have come to realize that although some diseases become milder, others might evolve with greater virulence. We now see nature more as an arms race between life forms deploying assorted genetic strategies.

A second aspect of this more modern viewpoint, is to realize that the scale on which we view events is important. Improving a species through evolution inevitably involves the death of many less fit individuals. Applying this Darwinian idea to human populations lets us see that whereas mass fatalities from a plague are tragedies at the personal level, they can have positive effects when seen from a long-term perspective.

These positive effects vary from genetic changes that make us more resistant to the disease responsible for the epidemic (and often to related infections), to effects on human society that are hard to pin down and quantify. Epidemics have undoubtedly affected the outcome of many wars and conflicts. Whether these interventions were a good thing obviously depends on which side you support. Less ambiguous is the contribution of epidemics to the development of a free, technologically based society in the West. More ambiguous are the possible effects on religious belief and human behavior.

Modern progress in DNA technology and human genetics is generating a vast amount of data. Analyzing and checking this will take time. The next few years should reveal many connections between infection, disease resistance, and alterations in genes that affect not only our physical characteristics, but also brain function or development and thus impact human behavior. We live in exciting times!

David Clark
Carbondale, Illinois

Acknowledgments

I would like to thank Donna Mueller for commenting on some early drafts, and Kirk Jensen for long-term editorial support through several versions of the manuscript.

About the Author

David Clark was born June 1952 in Croydon, a London suburb. After winning a scholarship to Christ's College, Cambridge, he received his Bachelor of Arts degree in 1973. In 1977, he earned his Ph.D. from Bristol University for work on antibiotic resistance. David then left England for postdoctoral research at Yale and then the University of Illinois. He joined the faculty of Southern Illinois University in 1981 and is now a professor in the Microbiology Department. In 1991, he visited Sheffield University, England, as a Royal Society Guest Research Fellow. The U.S. Department of Energy funded David's research into the genetics and regulation of bacterial fermentation from 1982 till 2007. David has published more than 70 articles in scientific journals and graduated more than 20 masters and Ph.D. students. He is unmarried and lives with two cats: Little George, who is orange, and Ralph, who is mostly black and eats cardboard. David is the author of *Molecular Biology Made Simple and Fun,* now in its third edition, as well as three more serious textbooks.

1

Introduction: our debt to disease

Early in the fifth century A.D., the Huns, led by Attila, emerged from the Asian steppes and swept across Europe. They faced no serious resistance. What was left of the greatest civilization the world had seen, the Roman Empire, was a tottering wreck. One more good shove and the remains of Roman civilization would have taken a final nose-dive. But strangely, on the verge of storming Rome itself, Attila withdrew. Why?

For many centuries the official answer was that God had intervened in some mysterious way to protect his chosen city, Rome, seat of the papacy. In more recent times, such supernatural explanations have fallen out of favor and the question has arisen anew. Some have suggested that Attila was overawed by the sanctity of Rome. But why would a pagan warlord like Attila stand in awe of a Christian center? Attila was by no means an ignorant barbarian: For example, he invited Roman and Greek engineers into Hun territory to install bathing facilities. However, his respect for Roman civilization was clearly of a pragmatic rather than a religious nature. Another theory is that Attila was worried about leaving unattended his newly acquired homeland, in what is now Hungary. But then why did he venture out almost as far as Rome and hang around indecisively for so long before returning? All these explanations founder on the same point. It seems clear that Attila did indeed set out with every intention of taking Rome, but his expedition came to a premature halt.

Mounting modern evidence suggests that Attila was stopped by a virulent epidemic of dysentery, or some similar disease. Most of his men were too ill to stay on their horses, and a significant number died. In short, bacteria saved Rome. The ancient world had no knowledge of bacteria. Instead, most ancient cultures believed that epidemics were one of the main ways the gods expressed their displeasure. In the Bible, pestilence is often a punishment for wickedness, both for disobedience by the Israelites themselves and for intrusions by outsiders. For example, an epidemic saved the holy city of Jerusalem from the Assyrian invaders, providing a precedent for the failure of Attila to take Rome. So, in a curious way, the earlier explanation of God preserving Rome has reemerged in a modern scientific guise.

But before we rush to enroll the bacteria as honorary Roman citizens, we must consider another aspect of the issue. A major reason Rome itself was in such disarray when Attila approached was that it, too, had fallen victim to pestilence. Several catastrophic epidemics had swept through Rome in the period before the Huns surged into Europe. So whose side were the microbes really on?

Nowadays, floods, earthquakes, and volcanic eruptions are regarded as "acts of God," at least by insurance companies. The implication is that neither the victims nor anyone else is responsible. This is not entirely true. People who persistently rebuild their homes on a flood plain or along a fault line are at least partly to blame. Similarly, epidemics do not just happen to anyone at anytime anywhere without good reason. Neither the epidemics that struck Rome nor the disaster that overcame Attila's Huns were just random outbreaks of disease. What's more, their origins were interrelated.

Before Attila, Rome had several narrow escapes from other hordes of barbarians. Several times it looked as if the end was near and that the Romans would be overwhelmed. Yet somehow the Romans scraped by. Part of the credit must go to the Romans, who were an unusually determined people, not prone to giving up easily. Yet much of the credit also belongs to the unseen and unsung legions of microbes. It is relatively easy for us today to understand why an overcrowded, unhygienic ancient city suffered from persistent outbreaks of pestilence. Why disease so often intervened to protect the same city from successive waves of barbarians is more difficult to understand.

Imagine an ancient society that is moving along the path to urbanization. Large numbers of people are crowding into a growing city, such as Babylon, Athens, or Rome, which is much larger than neighboring communities. Infections normally spread more efficiently through crowded cities than through sparsely populated villages and rural areas. Sooner or later, some pestilence or plague will strike the emerging city. Its population will be decimated, and for a while it will be vulnerable. But if it recovers, its population will consist largely of those who are resistant to the plague of the day. In other words, denser populations are the first to build up resistance to the current infectious diseases in their region of the world. Next time a major conflict arises, the movements of armies or of refugees will spread infection around the war zone. People from rural communities or smaller towns will have built up less resistance than the population of the city-state, so pestilence will fight on the side of the biggest city.

Once a major population center gains a significant lead over its competitors, the pestilence factor will make it extremely difficult to overthrow. This indeed is what happened to ancient Rome. A series of epidemics whose identities remain unknown devastated the Romans early in their history. Later, barbarians who ventured too close to Rome routinely succumbed to massive epidemics that had only mild effects on the Romans. As long as the Huns retained their nomadic lifestyle, they would have been little affected by epidemics. Even if an occasional marauder caught some infection from more settled and crowded regions, it was difficult for pestilence to spread among small, scattered groups of nomads. Once the Huns aggregated into a horde, under centralized leadership, the situation changed radically. On the one hand, they had little previous exposure to pestilence, so they lacked resistance. On the other hand they now formed a large, dense population, ripe for the spread of invading microorganisms. In a way, Attila's tragedy was the result of this vulnerable intermediate situation between nomadism and urbanization.

The general principle that *pestilence favors societies that have become resistant because of prior infection* has had a vast effect on human history. It has not only directed the growth and survival of the empires of the Old World, but it also was the major factor in European invaders' takeover of the American continent.

Epidemics select genetic alterations

Another result of ancient epidemics that experts have only recently come to understand is the accumulation of alterations in the human genome. Through the millennia, a never-ending stream of hostile microbes has attacked and decimated human populations. Each time a human population is devastated by infectious disease, genetic selection takes place. People carrying genetic alterations that confer resistance, even if only partially, have a greater chance of survival. Consequently, their descendents will make up a greater proportion of the surviving population.

The result of constant epidemics is that, over the ages, distinct acquired genetic changes now protect us against many individual infections. We still carry these modifications in our DNA sequences, and recent investigations are revealing a steady stream of such genetic alterations, many surprisingly recent. Thus, in many ways, we are what disease has made us.

Yet another convoluted twist of fate appears here. Several well-known hereditary defects turn out to be side effects of resistance to disease. For example, sickle cell anemia is the result of hereditary resistance to malaria, and cystic fibrosis is associated with resistance to intestinal diseases that cause diarrhea and dehydration. A single copy of the cystic fibrosis mutation reduces water loss, thus protecting against a range of diseases whose most dangerous effect is dehydration. Two copies of the cystic fibrosis mutation slow water movements in the lung too much. So one copy of the mutation protects against disease, and two copies of the same mutation cause a hereditary defect.

The case of cystic fibrosis is especially revealing. The cystic fibrosis mutation is unusually common in those of northwest European ancestry. Calculations based on mutation rates and population genetics suggest that these mutations arose shortly after the collapse of the Roman Empire. This collapse led to a massive loss of general hygiene, especially in the water supply. Doubtless waterborne intestinal diseases spread like wildfire, and eventually, mutations providing resistance accumulated.

Every cloud has a silver lining: our debt to disease

The way epidemics have intervened in history shows that disease is not just a uniformly negative matter. The outcome of an epidemic may be quite complex, especially over the long term. Whether we regard any particular outcome as "good" or "bad" depends partly whose side we are on and partly on the relative weight we give to short-term versus long-term effects. In this book, I point out the positive effects of epidemics. This is not because disease is beneficial overall, but because these less obvious beneficial side effects often are overlooked. If a virulent plague rages through society, the obvious response is to stop it by whatever means possible, not to sit around fantasizing about its effect on future centuries.

Not surprisingly, we normally think of infectious disease as our enemy. When a successful program of vaccination wipes out a blight such as smallpox, we feel no remorse that a unique life form has suffered extinction. When we learn that throughout the course of human history infectious disease has been responsible for more deaths than war, famine, or any other cause, this only confirms our viewpoint. Indeed, the victories we have achieved over infectious disease are among modern man's greatest triumphs. Today industrialized nations have largely brought infectious disease under control. Unlike our predecessors of only a century or two ago, nowadays we mostly die from heart disease and cancer. Our longer lives give us time to reflect on the other side of this issue, and

I argue that, paradoxically, we also owe a great debt to infectious disease.

This approach is not merely idle intellectual self-indulgence. Infections that still threaten us either tend to cause disease in a subtler manner, or else they remain dangerous for other complicated reasons. The classic modern-day example is AIDS. This disease does not actually kill directly. Instead, it damages the immune system, allowing other diseases, impotent by themselves, to gain a foothold. Perhaps it is time for humanity to also take a more indirect and subversive attitude.

Over the long term, a positive side to disease emerges. Granted, if large black swellings are appearing in your armpits and you're about to die of bubonic plague, you'll find it difficult to maintain an unbiased perspective. Nonetheless, although the Black Death epidemics that ravaged Europe in the Middle Ages were devastating at the time, they had beneficial effects on a more global and futuristic scale. They shook up the repressive feudal system and, in the long term, made a major contribution to the evolution of Western democracy.

On the negative side of the balance sheet, we have the millions who died painful deaths in the plague epidemics. On the positive side, we must not forget those other millions who would have died in misery and poverty if industrial democracy had been delayed significantly. In our horrified emotional reaction to epidemics, we normally forget this latter aspect. We do not know for sure how many children would have died in infancy each century if the feudal system had continued. However, if we compare the infant mortality of 30%–50% that prevailed before industrial democracy with the less than 1% infant mortality of today, we can clearly see that millions of innocent lives have indeed been saved.

On a more individual level, Charles Darwin probably caught Chagas's disease while on his famous voyage on the *Beagle* around South America and the Galapagos Islands. His resulting poor health kept Darwin at home for much of the rest of his life. Instead of wandering off on more expeditions to observe nature and collect specimens, he stayed put and pondered the origins of living things. This may well have played a major role in Darwin's compilation of the most influential book of the last few centuries, *The Origin of Species*.

As already remarked, whether such indirect effects are "good" or "bad" depends on your perspective. Should we consider the happiness of the individual, the benefit to a particular group, or the overall betterment of mankind? For that matter, how do we define the "betterment" of mankind? Whatever your outlook, the effects of infectious disease have been undeniably important in changing the course of history. Perhaps it is not too fanciful to think of "good" and "bad" diseases. Some diseases, like bubonic plague, may have had some beneficial long-term side effects on human society as a whole. Others, like sleeping sickness, have no positive aspect. Whatever our moral perspective, the effects of an epidemic on the overall fortune of a tribe, nation, or even a whole continent may be quite different than the immediate effects on the victims.

Crowding and culling

Pestilence has molded both our cultures and our genes over the long term. But before tackling these long-term effects, let's look at what actually happened to human civilizations when diseases struck. Over time, humans increased in numbers and gathered together in villages, towns, and then cities. Populations grew larger and denser as civilization progressed. Sooner or later, when people are crowded together, infectious disease takes the opportunity to spread itself around, too.

Let's focus on severe infections with high mortality rates (such as smallpox, Ebolavirus, or bubonic plague). Each time pestilence passes through society, a sizable fraction is wiped out. The survivors give birth to the next generation, and numbers gradually increase again. Once the population is dense enough, another epidemic strikes and the cycle repeats. Over many generations, genetic resistance develops, so virulent pestilences wane into childhood diseases and may eventually fade away completely. Meanwhile, novel infections emerge and spread, taking the place of yesterday's retired plagues.

A crucial point for understanding the long-term effects of an epidemic is that death is distributed neither equally nor at random. Infectious disease strikes harder at some segments of the population than others. Let's start with factors that are wholly or mostly biological in nature.

First, older people and young children are especially susceptible. This is because the human immune system is not fully developed in the very young and is beginning to fade away in the very old. Conversely, some individuals may be immune to certain infections. Nowadays, such immunity is mostly due to artificial vaccination. Before this was available, immunity was normally acquired the hard way by catching the disease and surviving.

Second is the phenomenon of genetic resistance to disease. In contrast to immunity, which is acquired during an individual's lifetime, genetically based resistance is inherited from one's ancestors. Individual people differ greatly in their inherent susceptibility to different diseases. When a population suffers from a dangerous disease, those with genetically based resistance survive more often than others. Consequently, inherited resistance gradually builds up over several generations. For example, the earliest reliable accounts of smallpox in Asia and Europe suggest that it was fatal 75% or more of the time. Yet over the next thousand years, the mortality rate fell to around 10%–30%. Then when smallpox was carried to the New World, the mortality rate among the American Indians was 75% or more. Thus, a virulent disease is vastly more devastating in a population that has never been previously exposed and has had no opportunity to build up resistance.

Many other factors affect our susceptibility to disease. These range from mainly biological to largely social in nature. For example, those who are poorly fed or live in bad housing and are cold, wet, and dirty are much more at risk than well-fed people who are dry, warm, and clean. Obviously, the closer people are crowded together, the easier it is for infectious disease to spread. These factors all lie on the interface where social conditions merge with biological effects.

Many of these factors can be lumped together, at least crudely. To put it bluntly, poor people are both more likely to become infected and also more likely to die if they are infected. Unfair as it may be, this is inevitably true in all real-life human societies. Today this is most clearly seen in the contrast between the industrial nations and the Third World. However, throughout recorded history prosperity and status have had their advantages. Even in societies of apes and baboons, higher-status animals tend to be better fed, a factor that helps them fight off many infections.

But how do you get to be rich or poor, high or low? To be sure, one way is to inherit money or social position (as distinct from genes) from your parents. But this ignores how your ancestors got rich to start with. Do competent, industrious, attractive, healthy, and brave people tend to rise through the ranks of society? Or is it the ruthless, greedy, cowardly, and corrupt who claw their way to the top? Whichever of these alternatives you espouse, for better or for worse, during a virulent epidemic, fewer people at the top will die than those at the bottom.

The message of this book

Humans typically labor under the illusion that they control their own destiny. However, I argue in this book that infectious disease has had a massive unrecognized effect on human history and culture. Moreover, constant epidemics with high death tolls have occurred throughout history and have selected major genetic alterations in the survivors. Modern DNA analysis, including the recent Human Genome Project, has revealed that alterations have occurred in certain individual genes. But many more changes remain to be discovered.

Some of these genetic alterations have mainly physical effects, but others may affect brain and behavior. For example, it is possible, though not fully proven, that genetic alterations that predispose humans to schizophrenia also protect against viral infections. As yet, the genes presumed responsible have not been identified.

Before dealing with these issues, we need to understand that many of our infectious diseases have emerged only very recently, after humans developed agriculture and settled into towns and cities (as discussed in Chapter 2, "Where Did Our

Diseases Come From?"). We also need to realize that a disease's mode of transmission can alter its virulence over a relatively short period of historical time (as discussed in Chapter 3, "Transmission, Overcrowding, and Virulence").

2

Where did our diseases come from?

Africa: homeland of mankind and malaria

Africa is the ancestral homeland of mankind. Our species originated there perhaps 100,000 years ago. From Africa, humans spread through the Middle East and around the world. Not surprisingly, most of our original diseases evolved alongside (or inside) their human hosts in Africa, so the human race grew up in constant contact with parasites such as trypanosomes, which cause sleeping sickness, and *Plasmodium*, which causes malaria.

Many diseases adapted to tropical conditions were left behind by those who migrated to colder regions. In particular, many parasites that need a warm climate failed to adapt to the temperate zones. Conditions inside the human body remain fairly constant. Consequently, the susceptibility of a disease agent to climate depends on how much time it spends outside the body between infections. Bacteria and viruses that are passed directly from person to person are affected little. Parasitic worms whose larvae develop in rivers or lakes before reinfecting human hosts are highly susceptible. Diseases that rely on insects to spread them are greatly affected by climate because their insect vectors often cannot survive colder winters. Thus, mankind left behind malaria, sleeping sickness, yellow fever, and many other insect-borne diseases when we emerged from the tropics.

These diseases are still a major problem in tropical regions. According to the World Health Organization (WHO), some 500 million clinically observed cases of malaria cause a little over a million deaths each year, the majority in Africa. Of these deaths, about half are of children younger than five years old. Recently, AIDS overtook malaria and tuberculosis to become the leading cause of death among the infectious diseases, with around three million deaths. (Diarrheal disease and respiratory infections still head the mortality tables, but these are each due to several different infectious agents.) Tuberculosis kills around 1.5 million victims per year, slightly more than malaria. However, these deaths result from about 10 million cases, far fewer than the 500 million cases of malaria. Relatively few malaria victims die outright. Instead, they suffer life-long debilitation, which not only lowers their productivity, but also makes them vulnerable to other infections.

How important was malaria?

Malaria is sometimes quoted as having killed more people than all the wars and all the plagues in recorded human history. But although infectious disease as a whole has undoubtedly killed far more people than warfare, little compelling evidence indicates that malaria has outperformed all other infectious diseases. Although malaria has taken a steady death toll in Africa and other tropical zones, it was absent on the American continent until European colonization. Furthermore, although malaria nibbled at the heels of Europe until recently by infesting marshlands, the dense urban populations of temperate Europe and Asia were fairly unaffected. Moreover, until the recent population explosion, the population density of Africa and other tropical regions where malaria is endemic was relatively sparse.

Even in Africa, the evolutionary homeland of both man and his earliest diseases, was malaria really the number one killer before the last few centuries? Today *Plasmodium falciparum* causes most lethal attacks of malaria, whereas the other three species of malaria cause relatively milder disease and are rarely lethal. Although *P. falciparum* is presently spreading from Africa around the tropical world, the sickle cell mutation that provides resistance is found only in Africans indigenous to regions harboring *P. falciparum* malaria. The fact that the sickle cell trait is so harmful by itself suggests that it is a recent, emergency, evolutionary adaptation. Over longer periods, we would expect the build-up of resistance with less deleterious side effects, as is the case for many other diseases, including the milder variants of malaria. This suggests that the malignant, falciparum form of malaria is of relatively recent origin and that, even in Africa, malaria was present in its milder forms for much of early human history.

Moreover, in precolonial Africa, many other diseases that have since been largely eradicated due to Western technology were still active. Relative to malaria, yellow fever may be trivial today, but in the early colonial period, sailors to tropical parts feared it at least as much. Again, among many West African tribes, smallpox, which has now been completely eradicated, was historically feared the most. Although malaria has survived the onslaught of modern technology better, this does not necessarily mean it was the major killer before other diseases were brought under control.

Our fellow travelers

Malaria is the best known example of a disease that has accompanied our species from its earliest beginnings and remains a major health problem. However, it is by no means the only disease to have accompanied us since our origin. Tuberculosis, herpes, and typhoid are other well-known examples. This raises the issue of how an infectious disease avoids getting left behind when its human victims consist only of small, scattered bands.

Consider first a "recent" disease such as measles. This humans-only disease is highly contagious and is spread from human to human without relying on any insects or other carriers. After recovery, humans gain immunity from measles. Consequently, measles faces the problem of constantly finding fresh victims. When measles has finished infecting all members of a small isolated tribe, it has nowhere to go. Thus, diseases such as measles cannot persist unless civilization provides a densely packed crowd of victims. Clearly, measles is not one of our original diseases; we consider its origins later.

One way for a disease to avoid the predicament of measles is to be shared among multiple animals. Malaria and sleeping sickness are examples of this approach. Another approach is to remain dormant inside a host until fresh victims are available. Herpes, caused by a virus, and typhoid and tuberculosis, caused by bacteria, have all taken this route. Viruses of the herpes family may lie quiescent in nerve cells for years until provoked by stress to emerge. They may then spread to new victims. Typhoid can hide in the gall bladder of human hosts who show no symptoms but are a constant source of infection to others. Tuberculosis, caused by *Mycobacterium tuberculosis,* hibernates in the lungs.

Human remains showing signs characteristic of tuberculosis (TB) have been found dating as early as the Neolithic period, when settled agricultural communities first appeared (9,000 B.C. onward). X-rays of Peruvian mummies dating to before the European conquest show signs of tuberculosis, implying that the American Indians brought TB with them when they crossed the Bering Straits some 10,000–15,000 years ago. Extraction of DNA characteristic of *Mycobacterium tuberculosis* from some of these mummies has confirmed that it really was tuberculosis. The signs of TB in the skull of a half-million-year-old *Homo erectus* from Turkey are vastly more ancient.

It was once thought that tuberculosis might have moved into the human population from cattle, which suffer from a closely related form of the disease. However, recent DNA analysis suggests the reverse—that we humans transmitted tuberculosis to cows after domestication.

Many human diseases originated in animals

Despite the new DNA evidence that exonerates the cow from spreading tuberculosis, most of our present infections probably did originate from other animals. It seems likely that prehistoric hunter-gatherers were relatively free of infectious diseases, compared with historical and present-day man. The unusual susceptibility of American Indians to most diseases brought across the Atlantic from the Old World argues that the indigenous people of the American continent had never been exposed to these diseases. This implies that these diseases emerged after the ancestral American Indians split off from their Asian relatives approximately 15,000 years ago. Because the migrating tribes evidently did not import them into America, it seems that smallpox, measles, and so forth must have been human diseases for less than 15,000 years—perhaps less, even, than that.

Before rushing forward, a word of caution is in order. We are fairly sure that malaria is an ancient disease. However, malaria was not present among American Indians before contact with the Old World was reestablished. The reason is that malaria is carried by mosquitoes, which failed to survive when humans migrated from the tropics into the colder regions of Asia. The Asian ancestors of the American Indians had therefore left malaria behind before they entered the American continent. When invoking New World susceptibility for the age of a disease, we must keep this factor in mind. Other tropical diseases that cannot persist in colder climates may also be ancient despite not being carried to the Americas.

Note also that, apart from the domestic dog, the American Indians lacked the domestic animals characteristic of the Old World. When the Bering Strait was crossed, cattle, sheep, horses, and pigs had not yet been domesticated by the tribes who made the crossing. Many human diseases have come from these animals. Because the first humans to colonize the Americas did not take these animals along, they could not have caught their diseases after migrating.

As humans expanded around the globe and populations grew ever denser, our species became a living paradise for infectious disease. No other large animal in the known history of our planet has provided such crowds of individuals, packed closely together, just waiting for some pestilence to move in and multiply. Over the ages, infectious diseases have migrated from their original hosts whenever they made contact with the human species. Animals that have the closest relationships with humans have been the source of most diseases. This includes not just the deliberately domesticated animals, but also the rats and mice that have taken up residence in and around human settlements. Even today, most human towns and cities are home to more rodents than humans.

Dense herds of domestic animals and fields of closely planted crops have provided similar opportunities for colonization by diseases from related wild animals and plants. Some of these, in turn, have moved on to humans. Today, as the remaining jungles and rain forests are being explored and exploited, they have yielded more novel diseases. Lassa fever, Hantavirus, and Ebolavirus have all made the jump to man, the ubiquitous host. Although mankind has done a good job of exterminating other larger animals, we have kept many of their diseases.

Recent diseases from animals

When Hippocrates compiled his treatise in the fifth century B.C., the ancient Greeks did not know of smallpox, measles, bubonic plague (an Asian disease), syphilis (an American import), yellow fever (from Africa), or leprosy. They were aware of herpes, typhoid and/or typhus, tetanus, amebic dysentery, rheumatic fever, chlamydia (both venereal and trachoma), and gonorrhea (or something very similar).

Zhouhou Beijifang, written by Ge Hong in fourth-century-A.D. China, lists malaria, erysipelas, typhoid, dysentery, and cholera. In contrast to the ancient Greeks, leprosy and smallpox were now known, implying that these appeared roughly 2,000 years ago. These records, together with a variety of other ancient accounts, suggest that many diseases we are familiar with today were absent in early historical times and have appeared only within the last thousand years or so.

Viral diseases, which colonized humans only after civilization provided sufficiently crowded victims, include mumps, measles, German measles, smallpox, polio, influenza, and even the common cold. At least our hunter-gatherer forefathers didn't have to worry about scaring away the game with violent sneezing! These viral diseases all have close relatives in various animals. After making the jump to humans, they adapted over the centuries to their new hosts and, in many cases, lost the ability to infect their original hosts. For several of these diseases, evidence suggests that they were originally more virulent and have become milder over the years.

For all these diseases, victims who recover become immune. Consequently, these viruses must all keep moving through a constant supply of new hosts. Flu and colds do return year after year and reinfect the same people, but each successive epidemic comes from a newly evolved strain of virus. Although you may be reinfected with such new strains, you remain permanently immune to variants of flu or cold viruses from previous years.

Smallpox is a good candidate for a very recent addition to humanity's panorama of parasites. It was still quite virulent in the twentieth century, even among Old World populations. Smallpox was once thought to be derived from cowpox. However, recent genetic analysis has shown that camelpox is its nearest relative, so transfer from camels to humans in the Middle East seems plausible.

Probably the earliest recorded smallpox epidemic is mentioned in the Koran, which is consistent with a Middle Eastern origin. The siege of Mecca by the Ethiopians in 569/570 A.D. was routed by this epidemic. A.D. 570 was the year

Mohammed was born, and Islam had not yet been founded. Nevertheless, Mecca was already the holy city of the Arabs, and the Ethiopian Christians were hoping to destroy Allah's sacred shrine, the Ka'aba. The Koran credits Allah with slaughtering the Ethiopians.

The Islamic expansion of the seventh and eighth centuries spread smallpox throughout the Mediterranean area— or perhaps we might better say that smallpox cleared the way for Islam to expand, much as it later cleared the way for the Spanish conquest of Central America. The Islamic Empire crossed the Straits of Gibraltar from North Africa to conquer Spain in 710. The Arabs were still in Spain when the Black Death pandemic of the mid-1300s occurred. This seems to have helped tip the situation in favor of the native Spanish kingdoms, and the Arabs were gradually expelled over the next couple centuries. Smallpox moved quickly. In the year 737 A.D., a great smallpox epidemic in Japan caused major depopulation.

Which diseases from which animals?

Although apes and monkeys are more closely related to humans, they have provided few diseases. Herded livestock and rodent pests are more frequent sources of human infections. This is not really surprising: The greater the population density of the animals a disease infects, the more opportunity that disease has to grow more virulent, to evolve new variants, and to spread. In addition to population density, another critical factor is intimacy. Cattle, sheep, and goats are grazing animals and live in fields separate from their human owners. In contrast, pigs and chickens are found in farmyards and have had much closer contact with humans. Dogs live closest of all, often sharing the house with humans and the uninvited mice and rats found in and around all human habitations. As a result, dogs, pigs, chickens, and mice have tended to pass on more infections than sheep, goats, and cattle.

Nonetheless, our cousins the apes and monkeys have provided us with a few infections. Amebic dysentery probably came from the Rhesus monkey, which lives in the forests of Asia. AIDS, from the Human Immunodeficiency Virus (HIV), comes from African monkeys via the chimpanzee. Some 30%–50% of green monkeys found in Africa today carry Simian Immunodeficiency Virus (SIV), a close relative of HIV. In contrast, green monkeys living in the Caribbean show no traces of infection. These monkeys were brought from Africa during the seventeenth and eighteenth centuries, indicating that the modern group of SIV/HIV viruses has emerged since then.

Not all human diseases come from other mammals. Birds are almost certainly responsible for influenza, which still circulates among pigs, people, chickens, and ducks. Even today, birds live in large flocks, and in earlier times, colossal swarms of waterfowl inhabited the wetlands of Europe and North America. As the herds and flocks of domestic animals grew in size, they picked up diseases from birds and eventually passed them on.

Who owns which disease?

Despite our self-centered outlook, very few diseases are restricted to humans alone. If no reservoir of infection exists among animals, curing, immunizing, or quarantining all human victims should drive such a disease to extinction. This has been done for smallpox. The WHO began the eradication program in 1966 and completed it in the 1970s. So far, smallpox is the only human disease to be totally eradicated in the wild, although ongoing attempts have targeted polio and the parasitic Guinea worm.

Total eradication is rarely possible because most diseases, even those that primarily infect humans, also infect other animals. Malaria, yellow fever, and many tropical diseases also infect monkeys and apes; bubonic plague and rabies can infect cats and dogs. Are malaria and rabies really human diseases, or are they animal diseases that humans sometimes have the misfortune to catch? Although the dividing line is somewhat arbitrary, we can distinguish diseases that circulate mostly among humans, animal diseases that are occasionally caught by humans, and shared diseases that routinely infect several host species.

Some animal infections are occasionally transmitted to humans by accident and are rarely, if ever, passed from person to person. Rabies is certainly infectious but is almost never acquired from being bitten by another person. Glanders is a disease of horses that is occasionally transmitted to humans, such as stable hands who come in close contact with infected animals. Anthrax is primarily a disease of cattle but may cause epidemics with high mortality among humans. Under natural conditions—that is, before civilization—these diseases were probably unknown among humans. Only after the domestication of cattle, horses, and dogs did their diseases come into close enough contact to jump the species boundary. Some of these animal diseases adapted to their human hosts and have become genuine human diseases. Others have remained primarily animal diseases and only sporadically infect humans.

Different diseases evolve at different rates. Generally, the fewer genes are involved, the more rapidly the diseases evolve. Thus, viruses evolve faster than bacteria, which, in turn, evolve faster than protozoa. Consequently, many virus diseases have evolved so quickly that distinct human diseases have appeared. Often these have diverged so far from their ancestors that they no longer infect even other animals. Good examples are smallpox, probably derived from camelpox, and measles, which is related to distemper, a disease of dogs and related carnivores. In contrast, bacteria evolve more slowly than viruses, and we still share most of our bacterial diseases with other animals, although there are often specialized, human-adapted variants. For example, "epidemic" typhus is a human-specific version of "murine" typhus, which infects both rodents and humans. Protozoa evolve more slowly than bacteria, and we still share our best-known protozoan disease, malaria, with other apes and monkeys.

Are new diseases virulent to start with?

We often get the impression that whenever a novel disease jumps the species boundary and infects humans for the first time, it is incredibly virulent, as with Lassa fever or Ebolavirus. This is an artifact of journalism. If half a dozen people in some out-of-the-way place contracted a novel but mild illness, it would probably not even be investigated, let alone hit the headlines. Microbes are constantly invading the human body. Few make it past the human defense system. If a novel infectious agent does survive, whether it causes a mild or lethal illness is largely a matter of chance.

Lassa and Ebola illustrate this well. Recent investigations have revealed that milder strains of Lassa fever and Ebolaviruses have been infecting humans many years before the official "discoveries." People who are infected with a virus are immune to further infection, and their blood contains circulating antibodies that are specific to the virus in question. Analysis of blood samples from inhabitants close to the Lassa River in Nigeria revealed frequent cases of people with antibodies to Lassa fever virus. Many had no recollection of any illness recognizable as Lassa fever; others remembered attacks of moderate to severe fever, often attributed to malaria. Earlier outbreaks of Lassa fever in Nigeria thus ranged in virulence from scarcely noticeable to moderate, with the latter usually diagnosed as "aberrant" cases of malaria. Lassa was recognized as a new disease only when a highly virulent version hit the headlines in 1969. Ebolavirus has behaved in much the same way, emerging officially in Zaire in 1976.

Over the long term, novel diseases may adapt to humans or may go extinct. Adaptation does not imply that the disease becomes mild—merely that it gains the ability to survive and multiply in humans. The level of virulence acquired depends on the mode of transmission and how plentiful, crowded, and unhygienic the human hosts are.

Measles and its relatives

Among Old World populations, measles is a relatively mild childhood disease, only rarely causing severe complications. It is transmitted from person to person, and once a person has measles, he becomes immune and the virus disappears from his body. For measles virus to stay in circulation, it must constantly find a new supply of victims to infect. In societies where essentially everyone catches, survives, and becomes immune to measles as children, the only new hosts for measles are newborn children.

Calculations indicate that a human population of 300,000–500,000 individuals in frequent contact with each other is necessary to provide new children at a sufficient rate to prevent measles from going extinct. Before approximately

1,000–500 B.C., there were no individual cities with populations over a quarter of a million. The first to appear were
Babylon, capital of the Babylonian Empire, and Ninevah,
capital of the Assyrian Empire, followed shortly by Europe's
first real city, Athens. Until these were available, measles
could not have been easily maintained as a human disease.

Before this period, diseases such as measles had to rely
on moving from town to village within a region of civilization
linked by road or river. Dense enough populations were
found in the Middle East, in the valleys of the Tigris and
Euphrates rivers (Mesopotamia), starting around 3,000 B.C.,
and later in the valleys of the Nile (Egypt), the Indus (North
India and Pakistan), and the Yellow River (Northern China).
Each of these areas had enough people to keep an epidemic
disease in circulation, as long as the separate communities
were in efficient contact. This varied considerably over the
centuries, and no doubt the spread of infectious disease fluctuated wildly as a result. Many diseases probably jumped
from animals to man during this early period, burned their
way through a few unfortunate towns in close contact, but
then failed to spread any further. Some animal diseases may
have jumped several times before becoming established in
the human population.

Where did measles come from? Probably from man's best
friend, the dog. Measles is a member of the distemper virus
family, the Morbilliviruses. Distemper viruses are typically
found in carnivores and include seal distemper and dolphin
distemper, in addition to canine distemper. Canine distemper
can also infect other carnivores and is found among sea lions
and hyenas. Humans probably caught measles from dogs a
couple thousand years ago, and it has since evolved into a
milder form.

Comparing DNA sequences shows that rinderpest of cattle is the closest present-day relative of measles. However, rinderpest is an extremely virulent disease that still causes mass mortality among cattle and related wild animals (buffalo, antelopes, giraffes, wildebeest). Viruses that spread directly from person to person, or between individuals of closely related species, usually evolve to lower virulence over time (as you will see in Chapter 3, "Transmission, Overcrowding, and Virulence"). Taking this into account, the most likely scenario is that dogs passed distemper to humans, where it evolved into measles. More recently, humans passed the virus to cattle, where it evolved into rinderpest but has not yet had time to drop significantly in virulence. In agreement with this scenario, the major rinderpest outbreaks in the 1990s showed lower death rates in domestic cattle but still caused 50% or more mortality in buffalo, kudu, and other game animals. This argues that rinderpest has partly adapted to domestic cattle and has spread from there to wild game animals, where it is still highly virulent.

Measles in ancient history

Can we pinpoint the origin of measles more accurately? Not really, but we can amuse ourselves guessing. In the *Iliad,* Homer describes how Apollo, who favored the Trojans, sent a plague on the Greek army besieging Troy. Although Apollo was the archer god, he was not the god of war; his arrows carried pestilence. He was also the god of medicine—what gods send, they can also remove, if asked politely.

> *Then he sat down apart from the ships and let fly an arrow: terrible was the twang of the silver bow. The mules he assailed first and the swift dogs, but then on*

the men themselves he let fly his stinging shafts, and struck; and constantly the pyres of the dead burned thick.

The *Iliad* was written around 700–800 B.C., several hundred years after the actual siege of Troy. In his commentary written some 400 years later still, Aristotle was puzzled why Apollo would attack the mules first. Knowing as we do today that canine distemper is closely related to both measles and rinderpest, we might argue that Homer reported what actually happened: an outbreak of an early distemper virus still shared by dogs, mules, and humans. If so, the Greeks probably caught it from the more densely populated city of Troy.

As human population density increased over the next few centuries, a strain of this shared morbillivirus may have adapted solely to humans. Voilà, measles! It is even conceivable that the Great Plague of Athens (430 B.C.) was caused by the ancestral, highly virulent form of measles. At that time, Athens had just become the only region of Europe densely populated enough to maintain a disease such as measles. (At this time, the city of Athens is estimated to have had around 250,000 people and Attica, the surrounding area, another 200,000 or so.)

The morbilliviruses include several other mild diseases, including mumps and the human parainfluenza viruses. As you might guess, parainfluenza viruses cause symptoms similar to but less severe than genuine influenza. Close relatives of these are known in birds, pigs, and monkeys. Take your choice. Successive waves of infection by a series of related viruses probably entered the human population from various animals. Those that were originally lethal have by now become milder, and their symptoms are so vague that only mumps still merits a name of its own.

Diseases from rodents

Rats and mice bear the same relationship to sheep and cows as weeds bear to wheat and maize. They are neither truly wild nor truly domestic. Even though we never intended to domesticate them, our rodent pests depend as much on humans for their homes and food as animals we have deliberately tamed. Rats live in attics, basements, sewers, storm drains, and any other underground tunnels. House mice live in cavities within walls, and field mice live in barns, haystacks, and cornfields. Rodents scavenge leftovers and steal grain, cheese, and other stored foods. Though rarely seen, rats usually outnumber people in a typical human city.

The mouse was one of the earliest domestic animals. The Natufian culture, found in the Middle East about 12,000 years ago, was transitional between hunter-gatherer and a settled agricultural way of life. The remains of domestic mice were found when excavating the Natufian layer at Jericho. Presumably, mice were attracted to the stored grain and moved in with the humans. Today rats and mice star in cartoon movies, and the most famous American of all time is a mouse whose only close competitors are a duck and a rabbit. In historical times, rodents were more often star performers in spreading pestilence.

Some diseases are spread directly by rodents, but more often insects act as intermediaries. The most famous example is the Black Death (bubonic plague), spread by fleas carried by rats. This is not a uniquely human disease, but is shared by many animals.

Typhus fever is carried by lice or fleas. Epidemic typhus has adapted for human-to-human transmission via the human body louse. It is closely related to murine typhus, which spreads among rodents or from rodents to humans via

fleas. The rodent disease is the ancestor of the more specialized and more virulent epidemic typhus. This specialized
human version of typhus has been responsible for massive
casualties among soldiers during military campaigns and
among the inmates of prisons and other institutions. Most of
Napoleon's casualties during his disastrous Russian campaign
fell prey to typhus. Where lice are free to roam over blankets,
clothes, and hair, typhus is not long in following.

Most diseases spread by rats and mice are actually carried
by the fleas, lice, or ticks that live on the rodents. However,
several cases of direct transmission are noteworthy. Some of
the emerging viral diseases, such as Hantavirus and Lassa
fever, are spread by contact with rodent droppings or urine.
These are both primarily diseases of wild rodents, not house
mice or city rats. Unless they establish themselves in the
domestic rodent population, they are unlikely to become a
serious health problem. For now, as with Ebolavirus, they are
a scary but rare threat, lurking on the fringes of civilization.

Leprosy is a relatively new disease

Most people are under the impression that tuberculosis and
leprosy, both the result of infection by *Mycobacteria*, are
ancient scourges of the human race. This may be true for
tuberculosis, but leprosy is a much younger disease and probably came from mice. Both diseases are spread by direct personal contact, so both are under evolutionary pressure to
adapt by becoming milder. Ancient traditions depict leprosy
as highly contagious and virulent, hence the need to segre-
gate lepers. In apparent disagreement with this, leprosy
seems rather difficult to catch nowadays, even with prolonged exposure. Only about 10% of the family members of a

leprosy victim catch the disease, despite being in close contact and also being genetically related.

Before dismissing the earlier views, let's consider them from an evolutionary perspective. Leprosy probably was more contagious in the past. Over time, many people sensitive to leprosy have undoubtedly died, so surviving humans are inherently more resistant. In addition, to promote its own distribution, the leprosy bacterium has probably become less virulent. Although leprosy is eventually fatal if untreated, this takes a long time, thus giving the bacteria longer to find a new host.

The disease known today as leprosy is caused by *Mycobacterium leprae.* It is sometimes called Hansen's disease, not merely to reduce the social stigma to its victims, but because assorted diseases causing skin lesions and disfigurement were lumped together with leprosy until fairly recently. True leprosy causes characteristic bone loss at the extremities and can be diagnosed unambiguously in skeletons dug up by archeologists. The earliest traces of leprosy found so far date to the second century B.C. in skeletons from an oasis in the Sinai desert. Thus, the leprosy mentioned in the Old Testament is not the same as our modern disease.

Where did modern leprosy come from? The culprit is probably the mouse, which harbors a severe disease caused by the related bacterium, *Mycobacterium lepraemurium.* Cats also suffer from a form of leprosy, although the disease rarely penetrates the internal organs and the skin lesions generally clear up naturally after a few months. Although mice get a much nastier disease, both cat and mouse leprosy are caused by the same bacterium. We may guess that around 200 B.C., *M. lepraemurium* made the jump into humans and then gradually degenerated into *Mycobacterium leprae,* which can no longer grow outside a living victim or reinfect mice.

What goes around comes around

Many human diseases, both infectious and noninfectious, are self-inflicted wounds. The present prevalence of obesity in the West is the result of overeating and underexercising. Cigarettes are responsible for lung cancer. In addition, most of our present infectious diseases are related in some way to the rise of human civilization and the domestication of livestock and, though unwanted, of rodents. In the next chapter, we ponder the spread of disease and the changes in virulence, much of which can also be blamed on human activity.

3

Transmission, overcrowding, and virulence

Virulence and the spread of disease

How a disease spreads greatly affects its impact on human society. Diseases that spread efficiently will clearly infect more people. Less obvious, but no less important, its transfer mechanism determines how virulent a disease may become.

We must tackle two widespread misconceptions. Both generalizations are half true, and scientific investigations have only recently discovered their flaws. The first is the idea that because diseases adapt to their hosts, they will inevitably become milder if we just wait long enough. Thus, syphilis was extremely virulent when first introduced into Europe but nowadays is much milder. Similarly, childhood diseases such as measles and mumps rarely do much real damage, although they were once much nastier. But this trend is not inevitable. Recent findings indicate that, under some circumstances, diseases change little in their virulence or even get worse. Moreover, some, like bubonic plague, appear to oscillate in virulence.

The second issue is the prevalence of infectious disease throughout history. We tend to think that the farther back we go in history, the dirtier and less hygienic people were, and so the higher the level of infectious disease. This is broadly true if we restrict ourselves to the last 1,000 years of Western civilization. However, if we consider the broader sweep of human history, the prevalence of infectious disease has fluctuated wildly. For example, only in the nineteenth century did Western civilization regain the level of hygiene that existed during the prime of the Roman Empire. Again, in very early times, before urbanization began, when humans were still few and far between, infectious disease was probably much less frequent.

Infectious and noninfectious disease

To understand how disease has affected our history, we must understand how infections are spread. Until recently, infectious diseases were lumped together with a variety of other ailments. Historical societies were often confused about their causes and, consequently, about what precautions to take to avoid them.

We can classify diseases according to how they are acquired. Wounds, bruises, and broken limbs are the result of accidents or deliberate violence. Ancient societies were well aware of the effects of a sword-thrust or a fall off a cliff. As with violence, poisoning can be deliberate or accidental. Early cultures certainly understood the idea of deliberate poisoning, although the victims were often misdiagnosed. For example, the symptoms of arsenic poisoning include vomiting and diarrhea, which superficially resemble the effects of certain intestinal infections. Accidental poisoning, especially on a large scale, was sometimes confused with infectious disease and other times blamed on witchcraft.

Hereditary diseases are the result of genetic defects passed on by one's parents. People born crippled were frequently viewed as victims of divine displeasure (usually directed against their parents), and those who exhibited strange behavior, such as epileptics, were often seen as possessed by spirits (evil or good, depending on their society's outlook). Nonetheless, genetic defects are rare compared to other causes of disease and probably caused little confusion, even though they were not understood until recently.

Cancers tend to occur later in life. They happen because of a build-up of genetic damage over the years in nonreproductive cells. These genetic defects are thus not passed on to the children, but instead are confined to the multiplying cancer cells within a single person. Cancer cells grow out of control and destroy the body to which they belong. Toxic chemicals, both natural and artificial, and ultraviolet radiation from the sun are responsible for many cancers. Other cancers happen because the body's own genetic machinery makes occasional mistakes. Cancers draw notice only when people live long enough for genetic damage to accumulate. Until recently, cancers were responsible for an insignificant fraction of human deaths. Death by heart disease, stroke, or old age is largely a modern luxury. Historical populations rarely lived long enough or ate well enough for their arteries to clog with fatty deposits.

Infectious diseases from microorganisms have caused most deaths by far throughout recorded human history. In this respect, our own age is peculiar. Thanks to modern technology, we mostly live long enough to worry about heart disease and cancer. But for most societies throughout history, most people met their end from infections caused by micro-organisms of some kind. This is still true for some Third World countries. Despite this, scientists have understood the nature of infectious disease only since the late 1800s.

Infectious disease is caused by invisible microorganisms

The cause of infectious disease—and even whether diseases were actually contagious and could be passed from person to person—has been hotly disputed over the ages. Only during the late 19th century could science begin to investigate the microorganisms that cause disease. In early times, infectious disease was often seen as punishment from the gods. Later, disease was blamed on such things as night air, marsh air, or other vapors. This attitude is well illustrated by the phrase "You'll catch your death of cold." As we now know, "colds" are caused by viruses, not exposure to low temperature. Nonetheless, poor nutrition, poor housing, and exposure to extremes of heat or cold weaken potential victims. Dirt may not literally breed disease as once thought, but lack of hygiene allows germs to survive and spread.

An intriguing aspect of historical beliefs about infectious disease is that the common folk were proved to be right in the long run, and the educated were mostly wrong. The priesthood pushed the idea that disease came from the gods. People were told to stop sinning and to pray for forgiveness, not waste time attempting to understand disease. Rationalist intellectuals put forward a range of theories based on factors such as diet, personality, climate, dirt, decay, and offensive odors of various sorts. Until the last century or two, most intellectuals rejected the idea that disease was contagious.

However, the behavior of the population-at-large suggests that ordinary people were aware that disease was often contagious. Avoiding contact with those infected by typhoid, plague, smallpox, and malaria was a sensible precaution. During the 1600s, the wealthier inhabitants of London kept an eye on the weekly "Bills of Mortality," much as we tune in to the weather report nowadays. These "bills" were lists of recent deaths and their causes. When the number of cases of something especially nasty, like plague or smallpox, rose higher than normal, the wealthy fled London for their country estates and left the poor to take their chances.

Why did the scientific establishment take so long to realize that diseases are transmitted from one victim to another? I believe two factors are at work. First, many diseases are not directly contagious. Thus, although malaria is spread from person to person, it is carried by mosquitoes, and a person cannot catch it through direct contact with a human sufferer. Bubonic plague is even more confusing. It can be spread from person to person, but it is usually transmitted by fleas. From a practical viewpoint, avoiding those infected is still a good strategy—you would be less likely to be bitten by the same flea or mosquito. From an intellectual viewpoint, the observed lack of direct transmission favored the various environmental theories. Second, the technology to actually see microorganisms is of relatively recent origin. Speculation about tiny invisible germs goes back to the Roman author Varro (116–26 B.C.), but demonstrating their existence requires more than mere words: It requires a microscope.

How infectious disease spreads

Different contagious diseases spread in different ways. We can subdivide these into three major mechanisms. Some diseases spread by direct person-to-person contact. Others spread indirectly via inanimate objects. Yet a third strategy is for insects or other intermediaries to carry the infectious agent. The way an infection spreads greatly affects whether it becomes milder over the ages, stays much the same, or gets more virulent.

Certain diseases require prolonged contact of an intimate nature to move from one person to another. These diseases are relatively hard to catch and can often be avoided by changing personal behavior. The sexually transmitted diseases (STDs) such as syphilis, AIDS, and gonorrhea illustrate this scenario. Strictly speaking, the transfer of body fluids is involved here. This is important from a practical viewpoint, because such infections can also be spread by improperly sterilized hypodermic needles. This occurs both among the intravenous drug users of the industrial world and in the clinics of Third World nations that lack money for disposable syringes.

Other diseases are spread by direct personal contact, but with less intimacy than for STDs. Many venereal diseases probably evolved from ancestors who infested the skin and body surface in a more general way. For example, the chlamydia that infect the genitalia are closely related to those causing the eye infection trachoma. The specialized sexual versions likely arose in historical times only as human populations became denser (see Chapter 7, "Venereal Disease and Sexual Behavior").

Some germs are transmitted by bodily contact or via non-living objects such as doorknobs, paper money, clothes, and bed linen. Highly contagious virus diseases such as colds, influenza, measles, and smallpox are typical of this group, although most of these can also be transferred through the air. Many infections are transmitted from person to person through the air by coughing or sneezing. This is known as droplet transmission, because the germs are carried in microscopic droplets of saliva, phlegm, or mucus. Many of these germs fail to survive if they dry out completely. Tuberculosis, influenza, and colds are familiar examples. As the nursery rhyme says:

> *I sneezed a sneeze into the air,*
> *It came to ground I know not where.*
> *But hard and cold were the looks of those,*
> *In whose vicinity I snooze.*

Infectious agents can also be taken in with food or drink. Poor hygiene may result in food or drinking water being contaminated with human or animal waste. Typically, such infections affect the gastrointestinal tract and include the many types of protozoa, bacteria, and viruses that number diarrhea among their symptoms. The purpose of diarrhea, from the germ's viewpoint, is to provide an exit mechanism from the body and to recontaminate the water supply. Examples of waterborne diseases include *Cryptosporidium* (a protozoan), cholera (a bacterium), and polio (a virus). Infections caught from food are often referred to as "food poisoning," despite resulting from bacteria or viruses instead of poisonous chemicals.

Diseases are often carried by insects such as mosquitoes and flies or by animals such as rats and mice. These are referred to as vectors. Sometimes multiple vectors are involved, such as in the spread of the Black Death by fleas carried by rats or typhus fever by ticks carried by rodents. Controlling vectors usually limits the spread of a disease far more effectively than treating infected humans. Insects and their relatives, the ticks and mites, are the most common vectors. However, other animals may act as vectors, as in the spread of rabies by bats and squirrels, or of West Nile virus by migrating birds. Plague and typhus normally rely on fleas and ticks to distribute them, although, under some circumstances, they can spread from person to person. Other diseases are obliged to spend part of their life cycles in a second host. Thus, malaria must pass from human to mosquito and back again to complete its developmental cycle.

Many diseases become milder with time

Let's consider the spread of a virulent virus like Ebola from the viewpoint of the virus. After infection, the victim will most likely die in a few days. Before the first victim dies, the virus must find another victim to infect. Clearly, the longer the first victim moves around, the greater the chances are of the virus making contact with someone else. If the virus incapacitates the first victim too quickly, it will undermine its own transmission. Consider, too, the spread of the virus from village to village. As long as the virus stays in the same village, where plenty of potential victims live close together, it can get away with killing fast. But what happens when the village has been wiped out? The virus must now find another population center. This requires an infected person who is still fit enough to travel. Over the long term, movement between population centers may matter more than how a disease spreads locally within a group of people.

Now consider two slightly different Ebolaviruses. One kills in a day or two. The second takes a whole week. Virus 1 may wipe out a whole village, but it will find it very difficult to transfer itself to the next village. Even if a dying victim staggers within sight of the next village, its people will probably not allow him in. During plague epidemics in medieval Europe, many villages and small towns stationed archers to intercept travelers. Anyone showing symptoms of plague was warned away and shot if they ventured too close. While lacking in sensitivity, such quarantine measures were effective, and many small villages escaped entirely from epidemics that decimated nearby towns.

By comparison, a less virulent Ebolavirus will spread much more effectively. Infected refugees fleeing an infected village may reach another center of population before symptoms appear. Thus, if we have a mixture of viruses, the milder forms will spread more effectively and, over time, will predominate. Many diseases appear to have done just this and have evolved to become milder. Examples include gonorrhea and syphilis (caused by bacteria), and measles, mumps, and influenza (caused by viruses). What unites these diseases is that all are transmitted directly from person to person.

Ebolavirus infects humans now and then after emerging from some animal host, probably bats. It wipes out a few people in close contact, and then the mini epidemic burns itself out. Much the same is true of Lassa fever and other highly virulent diseases that burst out of the jungle every so often. Although they give the press the opportunity to spur apocalyptic hand-wringing, they are unlikely to spread far without getting milder.

Crowding and virulence

Earlier thinking held that, given time, all diseases would adapt, to become no worse than measles and mumps. Virulent diseases were newcomers, not yet adapted to a state of biological détente with their human hosts. This viewpoint sees man and his infections in a perpetual cold war, with casualties due only to occasional misunderstandings. This wishful thinking has obvious marketing appeal and still frequently appears in books and articles that popularize biology.

This scenario ignores the ugly side of both evolution and human history. The inhabitants of our history books did not merely suffer from childhood diseases while their mothers read them stories about rabbits and mice dressed in human clothes. Until our own privileged age, most people died of infectious disease, much of which small rodents spread. The purpose of evolution is not to make life better for humans, nor even to produce a balanced ecosystem. Indeed, the very idea that evolution has some underlying moral purpose is basically religious. Evolution is simply a mechanism by which different living things compete using various genetic strategies. Those that propagate their own kind more effectively increase in numbers, and the less efficient go extinct. Mother Nature has no maternal instincts.

No absolute reason exists for why a disease should not remain virulent, nor why it should not get more virulent. Some do. Indeed, the same disease may fluctuate in virulence as conditions change. The critical issue is which factors promote decreased virulence and which promote increased virulence. The two main factors are overcrowding and transmission mode. Consider again two variants of the same disease, one mild and the other virulent. If humans are closely crowded, the virulent version has the advantage: There is no need for the patient to linger for several days to pass on the germs. As long as plenty of new victims are available nearby, the best strategy is for the disease to grow as fast as possible inside the original victims, generating more germs to infect more people. The slower, milder version of the disease will be left behind. Diseases tend to grow in virulence when their hosts are plentiful and crowded closely together. Conversely, diseases evolve with lesser virulence when their hosts are few and far between.

A highly virulent epidemic may wipe out a substantial portion of the human population. This decreases crowding, which, in turn, selects for a decrease in virulence. Ultimately, you might think, a balance will be struck and both the population density of the host and the virulence of the infectious agent will settle down to a gentlemanly compromise. This is the microbiological version of the famous "balance of nature" myth. But instead of reaching a state of stable equilibrium, periods of population growth generally alternate with devastating epidemics. Chinese records illustrate this effect. Between 37 A.D. and 1718 A.D., 234 outbreaks were severe enough to count as plagues—that's one every seven years. Although not every epidemic covered all of China, the frequency is impressive.

Bubonic plague provides a nice example of a disease whose virulence oscillated. Beginning in the mid-1300s, repeated epidemics of bubonic plague swept across Europe until the 1600s (later in some places). When plague first reached a town or city, the first few cases were usually mild and the victims recovered. Once within the crowded confines of a town, the plague became more virulent, often switching to its pneumonic form, which is spread through the air by coughing. Anyone who caught pneumonic plague could be dead within a day. From the germ's viewpoint, this is no problem, provided humans coughed germs over and infected another victim within this time. In a crowded medieval city, this was normally the case. Toward the end of an outbreak, most of the population either was dead or had recovered and become immune. Hence, the plague became milder again as the number of available victims became fewer and farther between. The mild forms then spread to the next city, and the cycle repeated. After a couple generations, the population recovered to where it could provide a sufficient supply of fresh victims, and the plague might revisit the original city.

Note the time scale. Microorganisms evolve so fast that they can change their minds—or, rather, their genes—during the course of an epidemic lasting less than a year. As illustrated by the Black Death in a single city, mutants with increased virulence may appear and spread in only a few weeks, and the reverse occurs toward the end of the outbreak. Thus, the virulence of a disease such as plague neither decreases nor increases; it oscillates. A major problem for the historian is that if a disease can change significantly in a year, how did it behave a hundred years ago? A thousand? Ten thousand?

Today the human population is exploding. In many Third World countries, this is exacerbated by poor hygiene. Consequently, we can expect diseases that are efficiently transmitted from person to person to become more virulent. In addition, more people need more food. The tendency is to plant larger areas with the same crop, to improve efficiency. However, such crowding makes crops more susceptible to epidemics, just as with humans. The best-known crop disaster was the Irish potato famine, which resulted from overreliance on a single crop. When a virulent strain of blight fungus wiped out the potatoes, the Irish had little left to eat. Infectious disease then followed in the footsteps of malnutrition. Starvation itself killed relatively few—most victims died of cholera, dysentery, or typhus fever. Thus, crop failures and malnutrition amplify the effects of infectious disease.

Vectors and virulence

Virulence may increase when a vector carries a disease. If a germ hitches a ride from one victim to another via mosquito, it matters little that the first victim is too sick to move.

Indeed, this may even work to the germ's advantage. Mosquitoes will be able to land and suck blood without the victim swatting them. Diseases that are carried from person to person by some other agency have little motivation to evolve mildness toward humans. Rather, they must avoid disabling their carriers. What happens to the human victims is less important. Malaria, sleeping sickness, typhus fever, yellow fever, and many other diseases are spread by insects, ticks, or lice. These diseases are dangerous and show few signs of getting milder. Indeed, the more virulent form of malaria, *Plasmodium falciparum,* is today spreading throughout the tropics and subtropics from its original focus in Africa.

The best way to control these diseases is to kill the vectors, thus interrupting transmission. Spraying insecticides such as DDT greatly reduced the incidence of malaria in many areas. Sadly, malaria is making a comeback in many parts of the Third World, due partly to insecticide-resistant mosquitoes and partly to complacency and political disintegration. Irrigation projects such as dams, reservoirs, and irrigation canals often work well in temperate climates. However, in tropical regions, they may backfire. They create large bodies of stationary water that are ideal breeding grounds for the mosquitoes that carry malaria, yellow fever, and other diseases. The slowly moving water of canals also provides a suitable habitat for water snails that carry the parasitic worms causing schistosomiasis (bilharzia). An example was the spread of schistosomiasis during the Senegal River Basin development in West Africa.

Waterborne diseases use the water itself as a vector. Such diseases can also increase in virulence. The disease relies on contaminated water instead of an insect to carry it from person to person. But the principle is much the same: The disease does not rely on human victims for dispersal. Contaminated water supply normally spreads dysentery, cholera, and many other infections that cause diarrhea. Rivers can carry germs in untreated sewage downstream and infect towns and villages hundreds of miles away.

Reservoirs and carriers of disease

A disease reservoir is a source of infection outside the human species. Reservoirs are usually animals in whom the infection is mild or even causes no disease. For example, bats are a reservoir for rabies and probably also for Ebolavirus.

A carrier is a human who is infected but does not become ill. Although carriers show no symptoms, they may transmit the disease to others. Even if all the susceptible human victims are dead or incapacitated due to a virulent infection, a few carriers may keep the infectious agent in circulation. Carriers may travel from one town to another, or they may stay where they are and keep the disease alive to emerge at some future time. Clearly, a disease that can rely on symptomless carriers or an animal reservoir is under less pressure to become milder.

Many diarrheal diseases cause symptoms in only a fraction of their human hosts. The proportion of symptomless carriers varies immensely. It may be more than half, as in *Cryptosporidium* or amebic dysentery, or very rare (about 2%–3%), as in typhoid. In most cases, the germs simply live in the intestines without causing disease. Intestinal diseases in which a large fraction of the population shows no symptoms are, by their nature, relatively mild, at least in most adults. The casualties from such diseases are mostly infants in poor countries. Malnutrition and lack of medical care make infantile diarrhea a major killer under such conditions.

A few special cases are known of germs that have adapted specifically to inhabit some tissue other than the intestines.

In such cases, the disease may remain much more virulent. In typhoid carriers, the bacteria inhabit the gall bladder, emerging now and then into the intestine. From there, they can reemerge into human society. *Salmonella typhi,* the agent of typhoid fever, is one of the most virulent infections spread by the contamination of food or water with human waste. It is also a specifically human disease, unlike many other varieties of *Salmonella,* which are shared with assorted animals. These less dangerous relatives have no special hiding place and must therefore refrain from killing their multiple hosts to stay in circulation.

Some viruses also lay low in specific tissues, biding their time. The best known are chickenpox and herpes. In fact, chickenpox *(Varicella)* is a member of the Herpesvirus family and is unrelated to the true Poxviruses (smallpox, cowpox, and so on). Several related variants of herpesvirus cause cold sores and genital herpes. Although the symptoms may be suppressed by treatment or vanish spontaneously, herpes never disappears completely. A few viruses remain hidden in a quiescent state. Symptoms may re-emerge under certain circumstances—if, say, the victim undergoes a period of stress. Chickenpox may also lie latent in nerve cells, re-emerging later in life as shingles, a painful skin rash. After reemerging, the virus may be passed on to others.

Development of genetic resistance to disease

"What does not kill me makes me stronger."
—Friedrich Nietzsche

On average, the healthier, faster zebras escape being eaten by the lion and survive to carry on the species. Disease, like large predators, preferentially carries off the young, the old, the weak and crippled, and the feeble-minded, together with those who have no friends, family, or allies to help them. Vulnerability is not merely a physical matter. Declining mental alertness may increase vulnerability to disease due to lack of appropriate behavior. From a Darwinian perspective, both predation and disease improve the species, often in a rather nonspecific manner, by selecting for healthy and vigorous individuals. In addition, more specific effects occur.

When a virulent epidemic rages through human populations, some survive and some die. In the days before vaccination, antibiotics, and modern medical technology, what decided who was fortunate and who was not? In addition to sheer luck, both social and biological factors affect the chances of catching a disease, as well as the likelihood of surviving if infected. We start with the strictly biological factors.

First, we must distinguish immunity from resistance. Both protect against infection, although in quite different ways. Immunity occurs within a single lifetime. It results from previous infection by the same disease, or one closely related. The immune system remembers, and when exposed again, it rapidly extinguishes the invader. This assumes that a person survived the first encounter with the disease. Vaccination is based on deliberately exposing people to mild or crippled variants of a disease. This prepares the immune system for meeting the real-life, dangerous version of the disease. Immunity may be full or partial. It may last a lifetime or just a few years. Immunity is not inherited and cannot be passed on to children.

Resistance is genetic. A person is born with it or not, and resistance operates the first time you are exposed to a disease. After a lethal epidemic has passed, the humans who are resistant will have survived. Some who are sensitive will also have survived, for assorted other reasons. Nonetheless, if the death rate is significant, the proportion of people carrying genes for resistance will increase. The survivors pass these resistance genes on to their children, and the next generation starts out with a higher proportion of resistant individuals. If the same disease returns, it will kill a substantial number of the sensitive individuals in the new generation, and the proportion of resistant individuals will go up again. After several recurrences, the majority of the human population will be resistant.

Both smallpox and bubonic plague illustrate the emergence of resistance among humans. The earliest smallpox epidemics recorded in Japan had a 70%–90% death rate. By the mid–twentieth century, although smallpox was still dangerous, the death rate had fallen to around 10%–20%. Bubonic plague shows a similar history. The first European outbreaks in the mid-1300s were highly lethal, and several successive epidemics over the following century reduced the population of Europe by two-thirds. The same disease returned in the 1660s. Despite being called the Great Plague of London, both the infection rates and the probability of death among those infected were much lower. London lost scarcely 10% of its population.

Some diseases go extinct

If a particular infection returns periodically, it will find fewer susceptible victims each time. Eventually, sensitive humans may become so rare that the disease cannot find enough victims to continue transmitting itself and it may go extinct. This has happened to several diseases. Although many ancient records are ambiguous or lack medical detail, others describe outbreaks of pestilence whose symptoms are no longer familiar. Many of the early plagues of Rome and China either no longer are with us or have evolved out of recognition. The Great Plague of Athens, described so graphically by Thucydides, is the classic example.

The mysterious English sweating sickness caused quite a stir in historical times but failed to survive. The sweating sickness appeared in London in 1485. It was probably brought by mercenary troops from France who helped Henry VII seize the throne from Richard III. Symptoms included the sudden onset of fever, headaches, and "great swetyng and stynkyng with rednesse of the face and all the body." Most victims recovered, although a significant minority died within a day or two. Fatalities were oddly erratic; in some communities, 30%–50% were killed, while in other towns, almost none of those taken sick actually died.

The English and Germans were susceptible, but the Scots, Welsh, and Irish (that is, those of Celtic lineage) were mostly not affected. Neither were the French, who, if truly guilty of bringing it to England, must have suffered from only a mild form of the disease. The worst epidemic, that of 1528–1529, spread to Germany, Holland, Scandinavia, Switzerland, Lithuania, Russia, and Poland, but ignored France and the rest of Southern Europe. This suggests a strong genetic element in susceptibility. Curiously, the English upper classes were hit harder than the common people. Two successive Lord Mayors of London died in the first epidemic of sweating sickness, and in 1517, Cardinal Wolsey fell seriously ill but survived. In all, five outbreaks of sweating sickness occurred over about half a century, and then the disease faded away. A similar but milder disease, the "Picardy sweats," appeared in France during the 18th century, supporting the idea of a French connection. No known disease today has these symptoms.

Milder germs or mutant people?

When a disease gets milder, what has really happened? Did the disease change, or did the humans? Germs may mutate to avoid killing their victims too quickly, in order to spread themselves around. Humans may become resistant because sensitive individuals die off. Both processes occur in real life. Syphilis became milder. Humans became resistant to measles. In many cases, such as with malaria or leprosy, both processes have occurred. Because we are embarrassed talking about death and dislike thinking of the millions of humans weeded out by influenza, measles, and smallpox, we tend to talk of a disease getting milder even when humans became resistant.

Consider two alternative approaches for a disease to avoid killing its victims too fast. One is for the disease to become genuinely milder and nonlethal. Alternatively, the disease may remain lethal but kill only slowly. This is probably what happened to leprosy. Historical accounts suggest that leprosy was once highly contagious and far more virulent. Today leprosy is difficult to catch and will still kill if untreated, but this takes many years. Both victims and disease have changed genetically over time. Many Europeans carry genes for resistance to infection by leprosy.

Today we have direct genetic evidence for human resistance to schistosome worms, malaria, tuberculosis, leprosy, typhoid/cholera, HIV (AIDS), hepatitis B, and hepatitis C. The great sensitivity of indigenous Americans, both North and South, to influenza, measles, smallpox, and other Old World diseases implies that, here, too, genetic resistance has evolved in the Old World populations.

Group survival involves more than individual resistance

When a human society shows an altered response to a disease that is passed from parent to child, several possibilities exist apart from genetics. Behavioral avoidance is any cultural change that leads to protection. People who use mosquito nets and wear insect repellent become "culturally resistant." No genetic changes have occurred, but cultural resistance can be "inherited," in the sense that customs are passed from one generation to another. Some social effects also have an underlying biological basis.

People often use the term *herd immunity* to refer to two distinct protection mechanisms. Here we use the terms *indirect immunity* and *herd resistance.* Indirect immunity occurs when an immune or resistant majority shields a minority of sensitive individuals from infection. Let's suppose that 90% of a human population is either immune or resistant to some particular infection. The other 10% will be protected because the disease will find it very difficult to transmit itself through the population. The minority of sensitive people are hiding in the biological shadow of those who cannot catch—or, therefore, transmit—the infection. In practice, immunizing 75% or so of a population often breaks the chain of infection well enough to protect the unvaccinated minority. The exact numbers depend on the nature of the disease and its transmission mechanism.

Group resistance is quite different and results from having a large population with plenty of genetic diversity. The population has many alternative versions of the genes that protect against infection. Some versions work better against one disease; other versions of the same genes work poorly against the first disease but act well against other infections. Different individuals carry different versions of these protective genes. Even if a totally new and highly virulent disease appears, a large, genetically diverse population will contain some individuals who are inherently resistant. Even during the worst outbreaks of Ebolavirus, around 10% of those infected survived. Even if most individuals die, the species will survive.

Thus, the species, viewed as a unit, may be resistant despite the fact that most individuals are sensitive. Life goes on. Note that we are not talking about "diversity" in the sense of artificially mixing individuals of different races to produce a politically correct community. Most local human populations have considerable internal genetic diversity, especially in the immune system. Despite hitting previously unexposed populations, the Black Death in Europe, smallpox in the New World, and Ebolavirus in Africa all had mortality rates of 60%–80% against "racially pure" local peoples. Contrast the introduction of myxomatosis to Australia. Myxomatosis is a lethal virus disease of rabbits that was released in Australia to control the rabbit population. The initial epidemic killed over 98% of the rabbits. These rabbits were the inbred descendents of just a handful of European rabbits that had colonized the Australian continent. The rabbit population had little genetic diversity, so the die-off was colossal.

The implications of resistance to infection

Over the ages, humans have developed resistance to many infections. Some of these diseases have gone extinct; others have evolved to survive. When a relatively resistant human population and its diseases meet a previously isolated and sensitive population, there are major repercussions. The devastation of American Indians, both North and South, by measles and smallpox introduced by Europeans is a classic example. To Europeans, measles is a mild childhood disease, and smallpox, though not mild, has a death rate of only 10%–20%. American Indians had never been exposed to either disease and, therefore, had no chance to evolve resistance. Consequently, they died in droves.

Many other examples are known in which disease has devastated one side (sometimes both) in human conflicts. Sometimes disease fights for the home team. The colonial takeover of Africa was hindered more by malaria, sleeping sickness, yellow fever, and other gruesome tropical diseases than by military resistance from stone-age warriors, despite the world-renowned bravery of such peoples as the Zulu. However, dense urban populations, who have been previously ravaged by some pestilence and have developed resistance, generally have the advantage. When they come into contact with less dense populations, on the fringes of civilization, the barbarians usually sicken and die. Unfair as it may seem, pestilence usually fights on the side of the Empire, evil or otherwise.

Although it is clearly beneficial to be resistant to disease, sometimes there is an unexpected price to pay. We are beginning to realize that certain hereditary diseases are the dark side of resistance to infectious disease. Sickle cell anemia is a by-product of resistance to malaria, and cystic fibrosis of resistance to diarrheal diseases, especially typhoid. To understand this, we must review the mechanism of inheritance. Humans, like all higher animals, have two copies of each gene, one inherited from their mother and the other from their father. Thus, if one copy is damaged by mutation, a back-up is present.

Resistance to disease often results from having one mutant copy of a gene that is defective in its original function. Children who inherit two defective copies, one from each parent, may suffer from a hereditary defect. Children with one good and one mutant (that is, resistant) copy will be resistant to the disease in question. Children who get two good copies of the gene will be healthy but still be susceptible to the disease. In practice, a balance is struck, depending on how common and how dangerous the disease is and how crippling the hereditary defect is.

Resistance to malaria via the sickle cell gene reduces the oxygen-carrying capacity of the blood. One good copy of this gene allows the blood to carry enough oxygen. Those with one good gene and one mutant gene are resistant to malaria. Those with two defective copies might, in theory, be even more resistant to malaria. Unfortunately, they do not live long enough to find out, because they suffer from sickle cell disease and their blood cannot carry enough oxygen.

We are the survivors of the frequent epidemics that have emerged in the relatively short time since humans began huddling together in overcrowded towns and cities. Consequently, unlike most wild animals, modern-day humans carry many dubious genetic alterations that have allowed us to muddle through the short-term crises of successive plagues. How has this affected our overall health? Have these changes affected our behavior, intelligence, or other mental abilities? The precise effects are mostly unknown, although we are beginning to see a few glimmers, thanks to modern genetic analysis. One fascinating recent link is between the prion protein, whose malformed version is responsible for mad cow disease, and the certain receptors in the brain. The healthy form of the prion protein appears to protect the NDMA receptors from overstimulation. In genetically engineered mice, extra NDMA receptors produce higher intelligence. Would changes in susceptibility to mad cow disease change our intelligence? Perhaps. We are the children of pestilence, held together by genetic jury-rigging.

Hunting and gathering

Early humans were hunter-gatherers. Men hunted game; women gathered roots, nuts, and fruit. Our ancestors roamed in small bands, rarely meeting other tribes. Early hunter-gatherers occupied prime land with plenty of large game. Today's few remaining hunter-gatherers inhabit marginal areas in jungles or semidesert. Thus, the early hunter-gatherers were probably better fed. On the other hand, they did not have the option of visiting a modern hospital if injured, nor of trading skins and furs for portable DVD players and candy bars. Nevertheless, with some reservations, today's hunter-gatherers are the best illustration we have of conditions before most of mankind settled into an agricultural way of life.

Patterns of infection vary greatly between hunter-gatherers and settled, agricultural societies. Two major factors are intertwined: low population size and high mobility. Ancient hunter-gatherers almost certainly had much less infectious disease than we have today. As already noted, before the growth of dense human populations, most of our epidemic diseases did not exist. Furthermore, small, mobile, and relatively isolated tribes would rarely have been infected by contact with other groups. Today's hunter-gatherers tend to catch most of the diseases current among the settled farmers who live nearby. Nevertheless, they are still far less likely to be infected with the parasitic worms and intestinal protozoa that are circulated by the droppings of domestic animals and by irrigation programs. Their lifestyle of roaming over dry plains also protects them from the malaria that is typically found in marshy, wet, and irrigated regions.

Life expectancy and developing civilization

Overall, early hunter-gatherers were probably healthier and better fed than the sedentary farmers who followed them. Before civilization, life expectancy was probably around 30–40 years. Women bore five or six children, and infant mortality was perhaps as low as 30%, with a fair number of children dying between infancy and adulthood. Although most deaths were caused by infections, accidental deaths were also probably frequent among hunter-gatherers. The agriculturists who followed were more civilized than hunter-gatherers, in the sense of having better technology. However, their stationary lifestyle made them more susceptible to infections, and as villages grew into towns and towns into cities, disease became progressively more of a burden. Despite having more food, early farmers often had poorly balanced diets, as they relied on just a few staple crops. Meat consumption was low, as domestic animals were too valuable to slaughter routinely, and only the rich ate meat regularly.

In early societies, outbreaks of infection from domestic animals were probably quite frequent. But most of these would have died out rather quickly, due to lack of sufficient people—and animals—to keep the infection in circulation. Only after populations of a third to half a million were available could such new infections adapt to humans and survive as specifically human diseases. Although cities go back to roughly 4000 B.C. in Sumer, they were originally too small to support continuous epidemic disease.

The first cities big enough to keep diseases like measles continuously supplied with new people to infect appeared in the Middle East, and somewhat later in Greece, in the period 1000–500 B.C. Nonetheless, before this, there were densely settled human populations in the valleys of the Nile in Egypt and of the Tigris and Euphrates in the Middle East. To what extent epidemic disease spread before 1000 B.C. depended on the amount of contact between the individual communities strung along the river banks.

Somewhat later than the Middle East, dense populations arose in India and China, and provided similar opportunities for animal diseases to migrate into humans. Eventually, the three major centers of Old World civilization made good enough contact for infectious diseases to move from one to the other. Between around 500 B.C. and roughly 500 A.D., the disease pools of the Old World merged. The latter centuries of this period saw great political turmoil throughout the Eurasian continent. In the West, the Roman Empire fell, and in the East, China fragmented into multiple states, many under foreign domination. A thousand years later, over a period of less than a century, the combined infectious burden of the Old World was carried to the American continent, where it caused colossal devastation.

Disease and manpower

The massive casualty rates due to infant mortality plus periodic epidemics meant that many ancient societies were limited in their economic growth or military expansion by shortage of manpower. This is nicely illustrated by the values the Franks placed on different members of society. As the Roman Empire faded, these Germanic tribesmen took over Gaul, whose modern name, France, commemorates them. Shortly after 500 A.D., the Frankish law code (the *Pactus Legis Salicae*) set the *wergild,* the fine paid for wrongful death, for various people. A freeborn Frank was valued at 200 *solidi.* (A *solidus* was worth roughly one cow.) A woman who had survived to childbearing age was worth 600 *solidi,* as much as a member of the king's retinue. However, girls too young to bear and women past childbearing were worth only the standard 200 *solidi.*

Throughout history, the bulk of the human population was poorly fed and lived short, squalid, thoroughly unhygienic lives. In better-than-usual periods of history, those who survived infancy might hope to make it to 40 years, on average. In truly miserable periods of history, such as the early Middle Ages, blighted by war, famine, and pestilence, the overall life expectancy may have sunk as low as 20 years. As emphasized by Richard Dawkins in his book *The Selfish Gene,* evolution is a mechanism for spreading genes. The purpose of life is not to provide idle luxury for our bodies, or even challenging problems for our minds, but merely to pass on our genetic information to future generations. The chicken is just a fancy machine for laying eggs. The massive toll of premature death throughout human history has selected the fitter and stronger. In particular, it has selected for those who carry genes conferring resistance to the infectious diseases that have been our biggest killers. From a Darwinian perspective, civilization spreads successful genes, especially those that combat infectious disease, not cultural achievements.

How do microorganisms become dangerous?

Now let's pick up the story from the microbial viewpoint. Infectious agents vary greatly in their ability to cause harm. Before discussing the "professional" diseases, we must not forget the opportunists. When a person is weakened by injury, exposure, or starvation, or if the immune system is malfunctioning, microbes that are otherwise harmless may cause disease. Such opportunistic diseases have received a lot of attention in connection with AIDS. Victims of AIDS suffer damage to their immune systems, hence the name acquired immunodeficiency syndrome. Death results not from AIDS itself, but from the opportunistic diseases that can invade only humans lacking immune defense. These may be normal inhabitants of our skin or guts that invade tissues where they cannot normally survive. Others are microorganisms that do not normally infect humans.

The existence of such opportunistic invaders tells us that many microbes are poised on the edge of invasion. They can degrade and live on human tissues. If they could survive counterattack by the immune system, they would be able to invade us. Such microorganisms can become dangerous to healthy humans in two ways. The first is by gradual accumulating mutations that allow them to survive in human tissues. The second is by acquiring a preassembled set of virulence factors.

Cells carry their genetic information written in genetic code along the famous DNA molecule. Every now and then, DNA molecules are damaged by various causes and the information they contain is altered—or, as biologists say, mutated. Radiation of various kinds, including ultraviolet rays from the sun, can cause mutations. So can certain chemicals, such as those in soot, tar, and cigarette smoke. However, about half the mutations occurring in nature are spontaneous mistakes. Every time a living cell divides into two, it must duplicate its DNA so that its descendants get a copy of its genes. This process is not perfect, and occasional errors creep in. About one gene in a million suffers a mutation every generation. Many mutations have little effect; others are lethal. Despite this, a steady stream of mutations accumulates in living creatures.

Viruses mutate much faster than bacteria. Higher organisms, including protozoa such as malaria, change slowest of all. This is not because of a scarcity of mutations, but on how well the mutations are tolerated. This, in turn, depends on the relative genetic complexity. Random mutations are less likely to totally cripple a simpler organism, so the fewer the number of genes, the more rapidly an organism can change yet remain functional. Higher organisms have approximately 10,000–50,000 genes, bacteria have 500–5,000 genes, and viruses have 3–1,000 genes. Consider two machines, one with just a few components and the other with many. The more parts that interact with each other, the less flexibility we have to alter any individual component. If we randomly change the shape of one part of a complex machine, this will probably clash with the operation of another component. If we randomly change one part of a simpler machine, this is less likely to cause conflict. For example, we could double the length of the handle or the blade of a bread knife and still be able to slice bread. But if we doubled the size of a randomly chosen component inside an automobile engine, it would probably immobilize the car. Thus, the fewer genes, the more likely mutations will be tolerated and the faster evolution may occur.

Diseases from worms, fungi, or protozoa have changed relatively little during the course of human history. It is no accident that ancient descriptions of malaria are the earliest records of an infectious disease whose symptoms are clearly recognizable today. Conversely, viral diseases change so rapidly that they tend to become unrecognizable over the ages. Bacteria are intermediate in their rate of change. Around 400 B.C., the ancient Greeks described many infectious diseases still identifiable today. Most of these are bacterial, but the only recognizable viral infection is herpes. Many ancient epidemics cannot be identified today, even when their symptoms were carefully recorded. These were probably viral infections that have changed beyond recognition.

Genes are normally made of DNA. For day-to-day operations, cells make working copies of their genes in the form of RNA, a molecule related to DNA. The original DNA copy is stored safely until the cell divides. RNA is less stable than DNA and is copied less accurately. Therefore, it is much more likely to mutate. Nonetheless, some viruses contain genes made of RNA. These viruses suffer a massive mutation rate that no living cell could survive. Even among viruses, only those with few genes (no more than a dozen or so) can tolerate being RNA based. Even so, many of the individual virus particles produced are defective. Despite these drawbacks, many successful and widespread viruses use RNA. The benefit to the virus is that constant alterations camouflage it from the immune system. For example, influenza and AIDS are both RNA viruses. Influenza mutates so fast that this year's flu vaccine will be useless against next year's flu. AIDS mutates even faster. The AIDS viruses inside a single patient vary significantly from one another. This makes treatment with drugs extremely difficult, as resistant virus mutants arise inside a single patient.

Packages of virulence genes are often mobile

Although bacteria evolve slower than viruses, every now and then, some previously harmless bacterium becomes a full-fledged killer overnight. This results from the transfer of blocks of genes between different bacteria. One bacterium may spend a thousand years gradually mutating a few genes to better invade the tissues of its host animal. Then suddenly, the whole package may be transferred to a different bacterium, perhaps a harmless inhabitant of a different animal, and a novel disease emerges almost instantaneously.

Bacterial cells carry their genes, typically 3,000 or so, on a single giant circular molecule of DNA, the bacterial chromosome. Mobile clusters of extra genes are often carried on smaller DNA circles, known as plasmids. These divide in two when bacteria divide, so the genes they carry are inherited just like the genes on the main chromosome. About half of all bacteria found in nature contain one or more plasmids. Many plasmids can move between bacteria. Although most move only between closely related bacteria, a few promiscuous plasmids can move across family boundaries.

The enteric family is a related group of bacteria that mostly live inside animals, in their digestive tracts. They are widespread, and most are harmless. However, virtually all may become dangerous if they get a plasmid carrying virulence genes. Thus, bubonic plague is caused by *Yersinia pestis*. Its relative *Yersinia enterocolitica* sometimes causes mild diarrhea. *Yersinia pestis* itself has three virulence plasmids, while its less impressive relative has just one.

Although these blocks of virulence genes move from cell to cell on plasmids, they may occasionally be inserted into the bacterial chromosome. From being an optional extra, the virulence genes have become permanent fixtures. Among the enteric bacteria, the best-known cases are found in

Salmonella. The typhoid bacterium, *Salmonella typhi,* has at least three integrated blocks of virulence genes, and its less dangerous relatives, which cause food poisoning or mild fevers, have one or two. In addition, some *Salmonella* also carry virulence plasmids.

Viruses, plasmids, and virulence

Plasmids are clusters of extra genes, and viruses are packages of genes that infect cells. Is there a relationship? Sometimes, undoubtedly. Some viruses that infect bacteria have two alternative lifestyles. They may act like a typical virus and destroy the bacteria. Alternatively, instead of killing the host cell, the viruses may take up residence as circles of DNA. In other words, they become plasmids and replicate in step with the host cell. Other bacterial viruses may take up residence by inserting their DNA into the bacterial chromosome.

The best-known enteric bacterium, *Escherichia coli,* is a normally harmless gut inhabitant that has become famous because of its star role in genetic engineering. However, it can harbor assorted plasmids and viruses carrying virulence genes. As a result, some truly nasty strains of *E. coli* have emerged. *E. coli* O157:H7 debuted on the world stage in January 1993. It appeared simultaneously in Seattle, Nevada, Idaho, and California in contaminated hamburgers shipped from a central supplier to outlets of a single fast-food chain. Bloody diarrhea was followed by kidney failure that was sometimes fatal in children. In addition to virulence factors on plasmids, *E. coli* O157:H7 has a resident virus that carries the gene for shigatoxin. This toxin damages the kidneys and makes *E. coli* O157:H7 life-threatening instead of merely obnoxious. Shigatoxin is named after *Shigella,* which causes bacterial dysentery. At some point, this virus presumably infected *Shigella* and picked up the shigatoxin gene before moving on to *E. coli.*

Thus, plasmids or viruses can carry virulence genes. Transfer between closely related bacteria is obviously easier, but given time, a cluster of genes can move one step at a time between unrelated bacteria. Detailed molecular analysis has shown that the clusters of virulence genes found in enteric bacteria did not originate in members of this family. Presumably, these virulence clusters evolved long ago in bacteria of some other family and have subsequently moved. Where they first arose is still unknown.

4

Water, sewers, and empires

Introduction: the importance of biology

It is a major contention of this book that infectious disease has greatly influenced historical events. In this chapter, we consider the effect of disease, especially waterborne infections, on three major civilizations: the ancient Egyptians, the Indus Valley civilization, and the Roman Empire. Orthodox history emphasizes the actions of leaders, the strategies of generals, the policies of governments, and so forth. Sometimes the subsurface layer of economics is put forward as an underlying cause. But economics depends on the availability of raw materials and natural resources. However competent your rulers are and enlightened your fiscal policy is, creating a thriving society in a desert is difficult—unless, of course, you strike oil! This brings us to the realm of biology. Oil and coal are largely biological in origin: the decayed remains of vegetation that died many millions of years before man walked on Earth.

Behind politics lies economics, and behind economics lies biology. So then, I argue, to fully understand both history and current events, we need to include the biological perspective. Although this book focuses on the effects of disease, other biological effects are also important—these include environmental degradation, climate change, overpopulation, and pollution. In addition, these factors can help promote the spread of disease.

Irrigation helps agriculture but spreads germs

The first cities grew up alongside rivers: the Tigris and Euphrates, the Nile, and the Yellow River. Irrigation led to increased crop yields, which led to higher population density. These ancient river-based cultures then came to dominate the surrounding areas.

Ancient Egypt is remarkable for the long periods its regimes survived. The imperial Chinese dynasties that originated in the valley of the Yellow River were similarly long-lived. Settled agricultural societies based on irrigation are noted historically both for long-term stability and for the subservience of their common people. Irrigation requires large-scale public works, which, in turn, requires organized mass labor, which requires an autocratic state to control the lower classes. This political explanation ignores the fact that needs are not fulfilled by magic. Infection supplies the missing mechanism. Irrigation is extremely effective in spreading infections among agricultural workers. Waterborne diseases included a wide range of bacteria, viruses, and parasitic worms.

The class system, water, and infection

Throughout history, human societies have tended to divide into classes or castes of varying rigidity. Even in the same nation, even if they share the same religion and speak the same language, individuals of higher status avoid mixing with those of lower status. Early societies, primitive tribes, and social apes such as baboons all have a hierarchy, although they lack rigid divisions. Class systems are not based on merely social forces. Although those forces might contribute to class segregation, deeper biological reasons exist.

Infectious disease has always hit the poor and lowly far worse than the prosperous and powerful. Avoiding members of lower and more disease-ridden groups was not merely a matter of status. In the eighteenth and nineteenth centuries, for an upper-class Englishman to associate too closely with the working classes would have doubled his risk of tuberculosis. Respectable people did not associate with their social inferiors or visit houses of ill repute. Both sexual and social respectability reinforce avoiding infection. Today the threat of infectious disease has greatly diminished in the industrial nations. And yes, now that it's safe, every political leader who wants to pick up some cheap popularity arranges to "mix with the people."

Infection might also help keep the lower orders in their proper place. Even nowadays, some 50% of the Egyptian population is infected with assorted parasitic worms. Their feeble performance in the Arab–Israeli wars has been attributed in part to the debilitation of Egyptian soldiers by disease. In ancient Egypt, the peasants who waded barefoot in the flooded fields would almost all have been infected by parasitic worms carried by water snails (schistosomiasis). The military aristocracy would have been far healthier, stronger, and more vigorous. In contrast, unsettled peoples, such as the Huns, Mongols, Vandals, and Arabs, tend to be difficult to rule and their empires short-lived. Here, the lower classes are not subject to debilitating disease as an inevitable result of their economic role.

The origin of diarrheal diseases

The human gut provides a home for colossal numbers of bacteria. In the small intestine, mixed in with the material being digested are several million bacteria per gram (1 oz = 28.4 g). In the large intestine, the numbers per gram can rise to 100,000 million, or more than 20 times the world's human population. The vast majority of these bacteria are harmless, and some are beneficial by aiding digestion, making vitamins, or defending their habitats against other, more dangerous intruding microorganisms.

As we have already seen, from time to time, one of our harmless microbial inhabitants goes bad and causes disease. In addition, harmful newcomers might originate from other animals. Most of these newcomers probably arrived only after denser city populations provided more opportunities. In primeval times, outbreaks of a novel infection would have been confined to one or a few bands of nomadic hunters, and the disease would not have survived for long. As human populations got denser, they provided more scope for rogue microorganisms to circulate further and live longer.

Viruses that live in the intestine follow much the same pattern. These viruses infect the cells of the gut lining and cause a small amount of damage. Most cause so little damage that they go virtually unnoticed. A few trigger outbreaks of diarrhea. Rotavirus is the most common virus that causes infant diarrhea. In third-world nations, it can be life-threatening, although in industrial nations it is rarely dangerous. Rotavirus has been with mankind long enough for special adaptations to evolve. The protein, lactadherin, is synthesized in the breast and found in human milk. Instead of being digested by babies, it remains in their intestines. Lactadherin mimics the receptor on intestinal cells to which rotavirus binds. When rotavirus binds to lactadherin, instead of attaching to the intestine, the virus remains floating in the gut and is flushed out. Infants whose mothers provide them with higher levels of lactadherin suffer much less from rotavirus-induced diarrhea.

Cholera comes from the Indian subcontinent

We can only speculate on the effects of infectious disease on the earliest civilizations. However, intestinal diseases were probably the first to topple whole human cultures. The first civilization whose collapse can be reasonably credited to infection is the Indus Valley culture, sited in what is now Pakistan and part of northern India. Good evidence indicates that cholera or a closely related waterborne infection might have been instrumental in its collapse. To understand this, we first need to consider the natural history of cholera itself.

Cholera is caused by the bacterium *Vibrio cholerae*. Its victims suffer severe diarrhea and can die of dehydration if not cared for. Cholera is thought to have origins in India and emerged onto the world stage only in the nineteenth century. Hindu physicians first described cholera around 400 B.C., but the disease was almost certainly endemic there much earlier. It has caused epidemics in India ever since, especially in the region of the Ganges delta. Hindus from all over India make pilgrimages to Benares, a holy city on the Ganges. Pilgrims in large numbers moving to and fro across India tend to spread waterborne disease between major population centers. This adds to the effect of the dense local populations in contaminating the Ganges River complex and the Bay of Bengal into which it empties.

Cholera stayed in India until the early 1800s. During the following century, half a dozen worldwide pandemics erupted. In 1817, cholera spread both westward to the Middle East and Southern Russia and eastward to Malaya, Thailand, and Japan. It reached England in 1831 and North America the next year. Another cholera pandemic was responsible for massive casualties during the Crimean War of 1854–1856. Of 250,000 troops engaged, the British and their allies lost approximately 20,000 to cholera.

But why did cholera not sweep through the unhygienic cities of Europe and Asia until the 1800s? European ships were in regular contact with India from the 1500s, and British troops and traders in India suffered substantial losses from cholera while in India. The answer is unknown, but a reasonably convincing suggestion is that cholera outbreaks burn themselves out in a few weeks, unless they can infect fresh victims. The slow-sailing ships of earlier times took a relatively long time to reach Europe, so cholera taken aboard in India could not survive the voyage. Faster steamships and the opening of the Suez Canal have since brought East and West into much speedier contact.

Cholera and the water supply

In 1832, John Snow investigated the cholera outbreak at Killingworth colliery near Newcastle in northern England. He concluded that cholera was not carried by bad air or passed directly from person to person. He blamed the constant diarrhea, unwashed hands, and shared food. The miners stayed in the pit for eight or nine hours, taking their food and drink with them. As one informant put it:

> *I fear that our colliers are no better than others as regards cleanliness. The pit is one huge privy, and of course the men always take their victuals with unwashed hands.*

Snow continued his detective work in the London slums. The poor, crowded together in dirty conditions, were easy targets for cholera. But how did the rich contract the disease? Snow realized that the communal water supply could be contaminated by infected sewage. This explained why some communities were hard hit, whereas others, close by but supplied by a different water main or drawing water from a different well, remained unscathed. In the cholera epidemic of 1851, there were 4,093 deaths among the 266,000 people who drank water supplied by the Southark and Vauxhall water company that got water from the sewage-laden River Thames. In contrast, there were only 461 deaths among the 173,000 people whose water was supplied by the Lambeth water company, whose sources were uncontaminated.

Snow demonstrated that cholera is spread by contamination of water with the copious diarrhea its victims generated. England, followed by the other industrial nations, then took action to keep its water supplies clean. By the beginning of the twentieth century, most European and North American cities were as clean as the Romans had been in the first century. That's progress! The last great cholera pandemic, which began in India in 1891, reached only as far as Russia. Western Europe went essentially unscathed.

The rise and fall of the Indus Valley civilization

The Indus Valley civilization flourished from about 3,000 B.C. to 2,000 B.C. and was centered around the Indus and Sarasvati Rivers in what is now part of northern India and Pakistan. By virtue of its location, the Indus Valley civilization was able to spread over a much larger area than the ancient civilizations of Egypt and the Middle East. Its two best-known cities were Harappa and Mohenjo-Daro, both now in southern Pakistan. Estimates of the population of Mohenjo-Daro range from around 30,000 to as high as 100,000. More relevant to infectious disease, several dozen towns of various sizes sprawled over the region, resulting in a large overall population, all in reasonable contact.

The Indus Valley towns were notable for being laid out in a well-designed grid. Streets were of fixed widths, 9m for main roads and 3m or 1.5m for lesser streets. Most houses were built to standardized designs with bricks of fixed sizes; the relative dimensions of virtually every brick used in the Indus Valley culture were 1 × 2 × 4. Perhaps the major achievement of the Indus Valley culture was its water supply and drainage system. All major centers had sophisticated communal plumbing, with water supply channels and drains. Almost every house in major centers such as Mohenjo-Daro had its own baths. Drains took the dirty water to a communal underground sewage system. The Great Bath of Mohenjo-Daro was the earliest public bath in the world. It measures 12m × 7m × 2.4m deep (approximately 40 ft. × 23 ft. × 8 ft.). Although no one really knows, most archeologists think that it served for religious purification, that it was a sort of baptism tank instead of a swimming pool. In any case, it was built of tightly fitting bricks that were plastered over and lined with a layer of bitumen, to make it waterproof.

Despite these initial technological advances, the Indus Valley culture remained stagnant, showing little progress beyond its initial achievements. Eventually, it collapsed, leaving no obvious successor culture. Around 1900 B.C., the Sarasvati River dried up, and its ancient course was rediscovered using satellite photos only in the 1990s. In contrast, several other minor rivers changed their courses to empty into the Indus River, which became overswollen and inundated its flood plain. By 1800 B.C., the Indus Valley culture had disappeared virtually without a trace, and urban civilization did not return to India for a millennium.

Were the changes in the rivers responsible? Some archeologists believe so. All the same, the cultures of Mesopotamia had to contend with floods, droughts, and the movement of the channels of the Tigris and Euphrates rivers. Although individual cities and regimes fell casualty, the culture as a whole did not collapse; the people relocated to other sites and continued evolving. The collapse of the Indus Valley culture has also been attributed to the arrival of the Indo-European (Aryan) invaders from the north. However, despite the usual exaggerated claims of antiquity typical for sacred writings, the Rig Veda, telling of the Aryan incursions, was assembled no earlier than 1,000 B.C. and was not written down for another thousand years. Thus, the invaders arrived long after Harappa and Mohenjo-Daro were already abandoned mounds. Furthermore, life in the countryside appears to have carried on with relatively little discontinuity during the collapse of the Indus Valley cities. Some archeologists talk of a "systems collapse," which means that they have no idea why it collapsed but find it embarrassing to admit this in front of laypeople.

Cities are vulnerable to waterborne diseases

After a millennium of highly organized urban life, cities scattered over hundreds of miles were all abandoned within a century or two. To explain the collapse of an entire civilization on such a scale, we need a reason why urban life became nonviable, yet rural life went on. The latest archeological levels of Harappa and Mohenjo-Daro revealed large numbers of unburied skeletons. Although this originally suggested invasion and conquest, closer examination of the skeletons has shown that they lack the characteristic marks typically left by swords, axes, and other weapons. This implies that the slaughter was not due to human agency. The only realistic alternative is an epidemic of some kind. The great strength of the Indus Valley civilization was also its Achilles' heel (despite the fact that Achilles would not be born for another 500 years!). A water-distribution system can also distribute waterborne disease. This is well illustrated by the 550 outbreaks of waterborne disease documented in the United States between 1946 and 1977. Although few were serious, these numbers illustrate the constant vigilance necessary to ensure a safe water supply.

City-dwellers in Harappa and Mohenjo-Daro got water from public wells whose rims were usually within a few inches of the ground. The drainage system was underground, though buried only a foot or two below the surface. Whenever drains backed up due to blockages or local flooding, there would have been massive contamination of the water supply. In fact, the system is so susceptible to contamination that early archeologists did not believe that the drains could have been used for human excrement. However, later excavations revealed latrines connected directly to the drainage system and the presence of wooden mesh baffles for screening out solid waste. Refuse heaps left by maintenance workers who entered the drains by manholes in the streets demonstrate that blockages were not uncommon.

Today we might wonder how they got away with this for a thousand years. The answer seems to be that highly virulent waterborne diseases had not yet evolved when the Indus Valley cities were first built. As we have already discussed, most virulent epidemic infections have emerged only in the last few thousand years. Before dense populations that shared the same water supply and sewers arose, severe diarrheal diseases had no way to spread. During the period from 3,000 B.C. to 2,000 B.C., bacteria that had previously relied on occasional mild diarrhea to wander from intestine to intestine were presented with ideal conditions for evolving into virulent waterborne killers. To put it rather unkindly, the evolution of cholera could well be the legacy of the Indus Valley culture. Cholera is perhaps the most likely agent, especially because it was known in India from early times. But because more than a thousand years passed between the fall of the Indus Valley culture and the earliest convincing descriptions of cholera, other diarrheal diseases, caused by related enteric bacteria such as *Shigella* or *Salmonella,* might have been the cause.

When a virulent form of cholera had evolved, it would have rapidly spread from city to city throughout the Indus Valley region. The older and larger cities would have been most susceptible, whereas rural communities would have suffered far less. Before modern medicine, epidemics of cholera typically killed a high proportion of those infected within a few days. When an epidemic struck, many city-dwellers fled to the countryside, probably under the impression that they had offended the gods. If you doubt that disease could cause the abandonment of entire cities, remember that, in medieval Europe, the death rate in the cities was greater than the birth rate, and the city populations were maintained only by continued immigration from the countryside.

Why didn't waterborne disease wipe out other urban cultures? Consider the Romans, who had an extensive system for water distribution. For one thing, the Roman sewer system was more effective in removing human waste. The Romans might have turned the River Tiber into a giant sewer, but the Romans did not generally use it for drinking water. Instead, aqueducts brought water supplies a considerable distance from elevated and sparsely populated regions. The aqueducts kept the water high off the ground, away from possible contamination. If anything, diseases that rely on contaminated water were kept at bay during the Roman period. Before moving on to consider what kind of diseases were the undoing of the world's greatest empire, let's reflect that the fall of Rome, and the collapse of its civil engineering system, provided the opportunity for diarrheal diseases to make a major comeback.

Cholera, typhoid, and cystic fibrosis

An interesting recent revelation is that cystic fibrosis is linked to cholera resistance. Cystic fibrosis is an inherited disease that affects about 1 in 2,000 white children but is rare in other races. You must inherit two defective copies of the gene, one from each parent, to suffer from the disease. Humans who have a single defective gene are carriers but do not show symptoms, because a single good copy of the gene is sufficient for normal health. The molecular basis of cystic fibrosis is a failure to secrete chloride ions across cell membranes. Water normally flows to follow the chloride ions. Deficient water flow means that the mucus that lines and protects the lungs is abnormally thick due to a lack of sufficient water diluting it. This not only obstructs the airways but also allows the growth of harmful bacteria. The bacteria are protected from the immune system by the mucus, which they also use as a source of nutrition. Cells lining the airways of the lungs are killed and replaced with fibrous scar tissue, hence the name of the disease. Eventually, the patient succumbs to respiratory failure.

In cholera, death results from dehydration. The mechanism involves the initial release of ions, including chloride, from intestinal cells. Water follows the ions, and diarrhea results. If one defective copy of the cystic fibrosis gene is present, chloride ions and water move more slowly, and this protects against water loss via diarrhea. Consequently, a single defective gene protects against cholera—or, for that matter, any other disease that causes diarrhea and dehydration—but does not inhibit water flow enough to cause cystic fibrosis.

About 4.3% of the white population carry one defective copy of the cystic fibrosis gene, and 0.05% have two defective copies. Until very recently, nobody who received two bad copies and, therefore, suffered from cystic fibrosis lived long enough to have children. This allows us to calculate the rate at which bad copies of the gene are eliminated from the gene pool. For the incidence of cystic fibrosis to stay constant, defective copies of the gene must be created by mutation at the same rate at which they are lost by selection. In practice, the required mutation rate turns out to be a hundred times higher than naturally occurring mutation rates. This tells us that the prevalence of cystic fibrosis cannot be due to the constant emergence of new mutations. On the contrary, the cystic fibrosis mutation, when present as a single copy, is somehow being favored and preferentially passed on. Deeper genetic analysis also shows linkage disequilibrium. This means that mutant versions of the cystic fibrosis gene are associated with particular versions of neighboring genes much more often than expected by chance. This confirms that the mutant version of the cystic fibrosis gene is being passed on in association with these nearby genes instead of being constantly re-created by new mutations appearing at random.

Many slightly different mutations at different sites within the cystic fibrosis gene can produce defects, and around 50 different defective versions of this gene have been detected. Nonetheless, 90% of defective cystic fibrosis genes found in Northern Europe have the same defect at the molecular level (the amino acid phenylalanine at position number 508 is missing). The other 10% include some 40 different defects. This indicates that the majority of defective genes have been inherited from a relatively small group of ancestors instead of appearing by random mutation. For the defective version of the cystic fibrosis gene to be maintained at its present level by inheritance, those who have a single bad copy can be calculated to have a preferential survival rate of 2.3% over those with two good copies.

But is this preferential survival really due to cholera resistance? If true, we would expect cystic fibrosis to be most common in Indians and other Asians, which is the opposite of what is found. As for the Europeans, if we assume a mutation rate of 1 in 100,000 per gene per generation, we would expect roughly 0.6% of the population to possess a single defective copy in the absence of any advantage. Cholera first appeared in Europe around 1820, approximately eight generations ago. An increase due to a selective advantage of 2.3% over eight generations would result in 0.72% having a single defective copy, a far cry from the 4.3% who actually have it. Assuming 20 years per generation, a selective advantage of 2.3% would need around 1,700 years to produce the level of 4.3% seen today.

The collapse of the Roman Empire and the ensuing decline in hygiene all over Europe fits the needed time scale rather well. This event would have resulted in the spread of many diseases, such as typhoid, bacterial dysentery, and rotavirus. Like cholera, these all share diarrhea as a symptom and are spread by contamination of water with human waste. Recently, the connection between typhoid and cystic fibrosis has been strengthened. It has been found that the typhoid bacterium uses the cystic fibrosis protein when it first enters cells of the gut lining on its way to invade the bloodstream. Thus, a defective cystic fibrosis protein specifically keeps typhoid at bay and generally reduces diarrhea. Typhoid, which has a fatality rate of 10% to 20%, was endemic throughout preindustrial Europe. Remember that infant mortality was well over 50% for much of this period, and perhaps half of these infant deaths were due to diseases causing dehydration via diarrhea. Survival of infants rather than adult casualties would have driven the selection for the cystic fibrosis mutation.

How did disease affect the rise of Rome?

It has often been suggested that malaria played a significant part in the fall of the world's greatest civilization, the Roman Empire. But before we deal with its fall, let's consider some biological factors involved in the empire's rise to greatness.

The Romans had two great virtues, discipline and pragmatism. The first contributed to the most effective military organization of ancient times. The second was expressed in their civil engineering. The Romans are remembered for their legions and their roads. But they also built a wide range of structures concerned with water: reservoirs, aqueducts, canals, irrigation channels, public baths, sewers, and many miles of water pipes.

Improved water engineering had two major biological effects. Better irrigation allowed the Romans to grow more food per acre and to support a greater population density. Moreover, by providing clean, fresh water and by building public baths and sewers, the Romans greatly reduced the impact of infections that are carried by contaminated water or human waste. Consequently, their population was healthier and they lost fewer victims to disease than their less hygienic competitors. More people meant more recruits for the legions. More soldiers meant more conquests and more land to irrigate. And so an upward spiral continued until other negative, biological factors limited it.

Increased population density is a biological factor that at first seems unfavorable. Consider a society that has progressed to crowding together a large number of people into growing cities that are significantly larger than neighboring communities, such as Babylon, Nineveh, Athens, or Rome. Any passing epidemic will spread more efficiently through the crowded city than in the towns and villages of more sparsely populated neighbors. Sooner or later, any such overcrowded city will be struck by pestilence. Its population will be decimated, and for a while it will be vulnerable. However, provided that it survives and recovers, its population will now consist of those who are resistant to the plague of the day. In other words, denser populations will be the first to evolve resistance to those infectious diseases that are current in their region of the world.

The next time a major conflict occurs, the movements of armies and of refugees will distribute any available infectious diseases around the war zone. The enemies of the big city-state, from smaller towns and villages, will have built up less resistance than the population of the largest city. Voilà! Infection will do most of the dirty work: The enemy will be devastated by whatever plague is in circulation, and the bigger city will gain a massive advantage. Cleanliness might be next to godliness, but the demons of disease fight on the side with the denser population.

But how can we have it both ways? First, not all diseases are spread by water or human waste, so improved hygiene from civil engineering still leaves plenty of scope for insect-borne scourges such as bubonic plague, typhus, or malaria, and for airborne diseases such as measles or smallpox. Second is the matter of temporal sequence—Rome was not built in a day. Early in its history, Rome was indeed struck by several epidemics whose nature is unknown to us because the records are fragmentary and give little clue to the symptoms. Though devastated, Rome pulled through.

Overall, a fine balance must be struck between the advantages of a disease-resistant population and a plentiful supply of healthy manpower. Consequently, great empires are few and far between. Nonetheless, when a major population center does gain a lead over its competitors, such biological factors make it extremely difficult to overthrow. Despite this, all good things eventually come to an end. As population density continues to grow, the pendulum swings back in favor of the spread of infectious disease. Sooner or later, the invisible legions of microorganisms make a comeback. In the case of Rome, this was much later and did not happen until the exhaustion of natural resources and resulting environmental damage gave infectious disease a helping hand.

How much did malaria contribute to the fall of Rome?

In ancient times, malaria was endemic in large areas of Italy, Greece, the Middle East, and North Africa, all of which were incorporated into the Roman Empire. Although we often think that ancient peoples had little idea of what caused disease, this is not entirely fair. Roman author Marcus Varro theorized that malaria was due to "tiny animalcules" that multiply in marshy places, float in the air, and sneak into people by the mouth and nose, causing fever and sickness. He was not wholly correct, yet he came close. Other Romans realized that the annoying insects that breed in marshes helped spread disease. Varro's real problem was not so much ignorance as having no way of proving his assertion. More than a millennium later, Western technology developed microscopes capable of seeing Varro's animalcules. And so those who thought that diseases were the result of God's displeasure, night air, or vapors from decaying garbage often dominated public opinion.

Did malaria really promote the downfall of the Roman Empire? Granted, the Roman Empire took a heavy toll from malaria, but so did all other ancient cultures inhabiting the coastal regions of the Mediterranean basin. Unlike bubonic plague or smallpox, malaria does not occur in epidemics that sweep through a civilization, leaving massive casualties over a relatively brief time span. Malaria is an endemic disease and takes up permanent residence in an area whose inhabitants are, therefore, subjected to continuously recurring infections. The disease takes a continual heavy toll not only on life, but also on vigor. In areas where malaria is endemic, the heaviest casualties are among the newborn. Adults quite often survive the disease but are seriously debilitated and frequently fall ill again from recurrent attacks.

Evidence from corpses reveals that malaria has been present in the eastern Mediterranean region since Neolithic times. Typical Egyptian mummies had their abdominal organs removed. However, mummification was developed gradually, and some of the earliest Egyptian mummies, which date to before the abdominal organs were removed, have large swollen spleens, symptomatic of malaria. Sometime in the fifth century B.C., malaria arrived in Greece, and by the next century, it was the most common serious disease. During the fourth century B.C., malaria moved into the marshlands near Rome. The Romans responded by instituting the worship of Dea Febris (Goddess of Fever). In addition, the Romans began recruiting soldiers for the legions from mountain areas free of malaria.

Three main ecological factors worked together to undermine classical civilization: deforestation, soil erosion, and the formation of marshlands. Heating, cooking, and smelting metal ores all consumed large amounts of wood. The grazing of livestock, especially goats, prevented the regrowth of shrubs and trees, making matters worse. Although they planted olive groves and fruit trees, the Greeks and Romans made little attempt to replace the forests used for fuel. Consequently, the growing population of the Roman Empire destroyed the bulk of the Mediterranean forests. This, in turn, led to massive erosion of the soil on the exposed slopes. As upland areas became less fertile, productivity fell. The soil swept downstream by rivers was deposited in more level areas, especially around lakes and at river mouths. There it formed flat, poorly drained marshlands that allowed mosquitoes to breed, thus spreading malaria. Although alluvial deposits provide good, rich farmland, the presence of malaria drove away many of the farmers. Those who stayed suffered constant debilitating attacks. Malaria is almost never totally eliminated from the body, and exertion can bring on subsequent attacks. Those who stayed either avoided hard work because it triggered renewed bouts of malaria or worked hard and fell victim to further attacks. Either way, agricultural productivity declined.

The refugees, who left the land, fled to the cities, where they relied on the grain handouts the government provided to the urban poor—the famous "bread and circuses." This problem was exacerbated by the land-reform policies of Roman populists such as Julius Caesar, who distributed land to retired legionaries with little experience in farming. By the time of the first emperors, Italy could no longer feed itself and Rome had become dependent on grain imported from the fertile Nile valley, which explains why controlling Egypt and North Africa was of such critical importance to the Romans. Of course, the densely populated valley of the Nile was also the channel by which diseases emerging from sub-Saharan Africa found their way into the Roman Empire.

Uncivilized humans and unidentified diseases

The rising numbers of urban poor and the consequent over-crowding provided opportunity for the spread of epidemic diseases. These were not long in making an appearance. Several major epidemics struck the Roman Empire in the first half of the first millennium A.D. The identities of these diseases are uncertain, until the Great Plague of Justinian (540 A.D.), which was almost certainly bubonic plague. The succession of epidemics gradually depleted the manpower needed for both the legions and the economy. Eventually, the empire became too depopulated to defend itself.

Why is it so hard to be sure of the identity of ancient epidemics? One issue is that ancient writers were often more worried about the effects of the pestilence and its religious implications than in accurate scientific diagnosis. Less obvious is that many diseases change over time or even go extinct, as discussed in Chapter 3, "Transmission, Overcrowding, and Virulence." Thus, even when described thoroughly, ancient plagues might be unrecognizable today. Despite this, historians often feel obliged to name the epidemic, almost always choosing a well-known modern disease.

For example, the first major epidemic to strike imperial Rome occurred in 79 A.D., just after the volcano Vesuvius erupted, and was largely confined to Italy. Guesses have included virulent malaria or anthrax. Neither makes any real sense. Recently, genuine evidence has come to light. Examination of bodies buried alive by the volcano has shown symptoms of brucellosis. Moreover, we know that the Romans used milk from sheep and goats without, of course, sterilizing it. Even today, various strains of *Brucella,* some virulent in humans, are found in sheep and goats and their milk and cheese.

The next was the plague of Orosius in 125 A.D. This started with a famine caused by locusts eating the crops in North Africa. The plague itself also began in North Africa and moved from there to Italy. The identity is uncertain. Whole villages and occasional towns were wiped out and abandoned.

The plague that started in the Middle East in 164 A.D. is named either after Antoninus, emperor when it began, or Galen the physician. Soldiers from Syria brought the disease back to Rome in 166 A.D. Corpses were removed from Rome by the cartload. The plague swept through the empire until 180 A.D. and, as a final blow, killed the emperor, Marcus Aurelius. After a brief respite, the plague returned in 189 A.D. This epidemic was the first to cause a break in the Roman defense perimeter. Before this, the empire continuously expanded and was able to hold its frontiers. In 161 A.D., a horde of Germanic barbarians, the Marcomanni, left Bohemia (now in the Czech Republic) and assaulted Italy from the northeast. Disruption from the epidemic left the Romans incapable of counterattacking until 169 A.D. Reports of the conflict suggest that most of the dead Marcomanni were actually killed by disease spread by the Roman legions.

The most famous physician of Roman times, Galen, fled from Rome during this plague. He also left a description of its symptoms. High fever, inflammation of the mouth, and diarrhea were followed by eruptions on the skin, although many died before this stage. One unproven and unlikely theory is that this was the first smallpox epidemic to hit the West. This hypothesis suggests that a smallpox outbreak in Mongolia set the Huns in motion. The Huns then both infected and displaced various Germanic tribes, who, in turn, infected the Romans. However, the Marcomanni apparently caught the plague from the Romans, not vice versa. Moreover, if the Huns and Germans had been decimated first, it is hard to see them applying much serious military pressure on their neighbors. Furthermore, the course of the epidemic does not resemble later, better-known European smallpox epidemics.

Next came the great plague of Cyprian in 250 A.D. Cyprian was the bishop of Carthage, in North Africa. He described violent diarrhea, vomiting, fever, ulcerated sore throat, and gangrene of the extremities. No rash or skin eruptions were noted, and the identity of the disease remains obscure. This was a true pandemic, spreading from Africa throughout the known world and lasting for some 16 years. It moved rapidly, both by person-to-person contact and on clothes or personal articles used by its victims. It was more virulent than previously recorded diseases, killing more than half of those who were infected. Panic followed pestilence, and refugees fleeing their homes spread the plague. Large areas of Italy were left uncultivated, and the empire was weakened by loss of manpower. By 275 A.D., the empire had retreated to the Rhine and the Danube, abandoning Transylvania and the Black Forest region. The emperor, Aurelian, took the unprecedented step of fortifying Rome itself.

Over the next couple centuries, successive epidemics, probably of the same disease, ravaged the Roman Empire. Barbarian attacks intensified, especially from the Goths and Vandals. A downward spiral of pestilence, famine, and war led to the decline and collapse of the Western part of the empire. Accurate records became few and far between as civilization fell apart. The collapse took longer than might have been expected because the Romans infected the incoming barbarians, whose hordes were thinned out by pestilence, too. In 447 A.D., Attila the Hun was approaching Byzantium, capital of the Eastern Roman Empire, when pestilence "of the bowels" (perhaps dysentery of some sort) broke out among his army. The Hun army was not destroyed, but the campaign was abandoned. In 452 A.D., as Attila approached Rome, there was a repeat performance and the Huns were halted by what was presumably the same disease.

Sometimes disease struck the barbarians after they had defeated the Romans. For example, the Vandals, who had taken control of Rome's northern African territories, were so devastated by a plague in 480 A.D. that they were swept away by the Moors, a nomadic Arab people. In 539 A.D., the Goths and Byzantines were fighting for control of Italy when the Franks burst in, hoping to take advantage of the confusion. According to the Byzantine chronicler Procopius, the Franks succumbed to the Italian secret weapons: dysentery and diarrhea.

For those who like economic theories, we should point out that the two great plagues of 164 A.D. and 250 A.D. led to the collapse of the Roman fiscal system. As is usual, casualties from the epidemics were heaviest among the poor. The ratio of peasants and laborers declined relative to the upper classes. The colossal die-offs thus eroded the tax base. In an attempt to maintain public spending on roads, irrigation, and other public works, as well as pay the legions, taxation rates were increased. This led to poverty and destitution among the surviving lower orders. Malnutrition and poorer housing resulted, which raised susceptibility to infection. A downward spiral of overtaxation, epidemic infection, and underpopulation thus set in.

Bubonic plague makes an appearance

In the East, the Roman Empire developed into the Byzantine Empire, based on Byzantium (Constantinople). The Byzantines fantasized about retaking the Western territories, especially Rome itself. The emperor Justinian (527–565 A.D.) came closest. After successful wars on his other borders, he invaded the West in 532 A.D. He retook North Africa, Sicily, Italy, and even part of Spain. If merely human enemies had opposed him, Justinian would have probably succeeded. But just as he was preparing to invade Gaul, another foe emerged: bubonic plague.

Vague accounts suggest that bubonic plague might have afflicted the Egyptians, Philistines, and other Middle Eastern nations since 1,000 B.C. or earlier. However, Justinian's plague was the first outbreak of bubonic plague described in sufficient detail that we are sure of its identity. To the physicians of Byzantium, it was a novel and terrifying disease.

Desperate to understand its cause, they performed autopsies on some of the victims. They found what they called "anthraka," the hardened remains of lymph nodes. Our word *anthracite*, a type of coal, comes from the same root and reminds us that bubonic plague was called the "Black Death" because the swellings turned into hard black lumps.

Fever was followed by the appearance of buboes, black swellings in the groin and armpits due to swollen lymph nodes. Death typically occurred on the fifth day, sometimes sooner and sometimes later. Procopius, archivist to Justinian, correctly records that bubonic plague was not directly contagious and that outbreaks began on the coast. Today we know that bubonic plague is carried by fleas, which, in turn, are carried by rats. The rats spread from country to country by ship, so the plague spreads from the ports inland. Because infection is by fleabite, those who came in direct contact with plague victims were no more likely to be infected than others within range of the fleas.

The plague of Justinian began in 540 A.D. in Egypt. It spread through the Middle East and, from there, to the rest of the known world. It struck Byzantium itself in 542. As is typical for bubonic plague, the mortality was low at first and then rose steeply. The inhabitants of Byzantium died faster than graves could be dug for them. The towers of fortresses were filled with corpses left to rot, and ships were loaded with bodies and abandoned at sea. The plague circulated till around 590 A.D. Many villages and towns were depopulated. The Moors retook North Africa, the Goths retook Italy, the Persians sacked Antioch, and the Huns nearly took Byzantium itself. The population losses from Justinian's plague took some 200 years to recover. During this period, the Islamic Empire established itself. The Byzantines and Arabs first clashed in the late 620s, while the Prophet Mohammed was still alive. After Mohammed's death in 632, fighting continued until 718, by which time the Islamic Empire had stripped Byzantium of the Middle East, North Africa, and Cyprus. The Byzantines believed this was divine retribution for the sins of the Christians. Doubtless the Moslems agreed!

The empire was at a serious disadvantage compared to the nomadic barbarians. The citizens of the empire were crowded closer together into towns and cities. In addition, sewers and drains provided the rats with ideal channels of distribution within every major town. Although their assembled armies were susceptible to infection, the barbarian population was spread thinly over more rural areas. Moreover, many barbarians were nomadic, moving frequently and living in wagons or tents. These provided far less opportunity for rats than permanent towns and cities. Thus, the empire ran out of recruits, while the barbarians still had plenty of men to draw upon.

By the time of Justinian's death, Rome was little more than a ghost town in the middle of malaria-infected marshland. The darkness had fallen. Not until the nineteenth century would civil engineering provide hygiene as good as that of imperial Rome. Even after the empire had retreated from many outlying areas, pestilence continued to take its toll. Thus, Roman-style culture continued locally in Britain for a hundred years or more after the empire pulled out. Continuing outbreaks of bubonic plague starting in the mid–fifth century and extending over the next hundred years were a major factor in the collapse of this and other briefly lived successor cultures.

5

Meat and vegetables

Eating is hazardous to your health

Smoking, drinking, eating, gambling, skiing, swimming, jogging, sitting still, playing computer games, watching TV, having sex, and taking narcotics are all hazardous to our health—although we usually are lectured only about those that aren't respectable. Despite the exaggerations of the anti-smoking lobby, most deaths in industrial nations are associated with being overweight from eating too much and exercising too little. Here we worry not about overindulgence, but rather the safety of our food supply. In particular, our focus is on the contamination of food and water by microorganisms.

Contaminated meat, especially processed meat such as sausage and hamburger, is the major cause of food poisoning in today's industrial nations. Food poisoning is usually the result of bacteria. Although bacteria might create toxins within the food, more often the symptoms come from swallowing the bacteria themselves, which then infect the gut.

Although *Salmonella* and *E. coli* are best known, several other bacteria, such as *Listeria,* and several viruses also contribute. Several massive recalls of frozen meat harboring *E. coli* O157:H7 have occurred. In 1997, the Hudson Foods plant in Columbus, Nebraska, was shut down and 25 million pounds of ground beef were recalled. This record was broken in 2002 when Pilgrim's Pride, America's second-largest poultry producer, recalled 27.4 million pounds of turkey and chicken products because of contamination by *Listeria.*

The number of outbreaks of food poisoning has been rising recently, especially in the United States. An estimated 80 million cases of foodborne disease occur in the United States each year, although less than 1% are normally reported. With a population of around 300 million, this means that roughly one in every four people suffers food poisoning in a typical year.

Processed meat can be safely sterilized by radiation, an approach that is widespread in Europe. A major problem in the United States is fear of radiation. Nations that do use radiation have far fewer cases of food poisoning. Another factor is the ever-increasing centralization of food processing. If one cow in a thousand carries *Salmonella* and the meat is sold by local butchers, only a handful of people will get sick. But if thousands of cows are processed in a central facility and their meat is mixed together in a huge vat, all of the meat becomes contaminated. A 2001 survey of ground meat (chicken, turkey, pork, and beef) in American supermarkets revealed that about 20% of the packages contained *Salmonella,* and over half of the bacteria were resistant to at least three antibiotics.

Although meat still leads, vegetables and nuts are increasingly a hazard. *Norovirus, Salmonella,* and *E. coli* are all involved. Spinach carrying *E. coli* hit the headlines in 2006, and tomatoes or peppers with *Salmonella* followed in 2008. In the latter case, the FDA first blamed contaminated tomatoes for a 30-state outbreak of *Salmonella* Saintpaul and later shifted the guilt to Serrano peppers, thus causing widespread confusion. The origin of this *Salmonella* outbreak was never resolved. More than 2 million pounds of pistachios were recalled in 2008–2009 because of *Salmonella* contamination.

Perhaps as much as 25% of fresh vegetables in the United States are contaminated with *E. coli*. Although these are mostly harmless, their natural habitat is the human intestine. Their presence shows that many fresh vegetables are contaminated by human waste. Contamination can occur during harvest or transport, or, perhaps more often, by irrigation with water that has come into contact with human waste.

The increase in food poisoning cases led both the U.S. Department of Agriculture (USDA) and the Food and Drug Administration (FDA) to demand more regulations. In the early 1990s, government meat inspectors still examined meat by eye, and standards of hygiene continued to deteriorate. The deaths of several children in the mid-1990s triggered long-overdue technical improvements. DNA-based screening now allows rapid and accurate tracing of contaminating microorganisms. This has led to much faster intervention, to cut off contaminated food or water at their source. Coupled with improved safety procedures, this has dramatically reduced food poisoning. Between 1996 and 2004, most foodborne illnesses in the United States dropped by up to 40%, except for *Salmonella*. However, since 2004, there has been no significant improvement, and *Salmonella* infections have increased slightly. Some 5,000 Americans still die of foodborne disease each year.

Hygiene in the home

"The intimate connection between a woman and a broom-handle is an obvious and natural fact."
—Suellen Hoy, *Chasing Dirt*

The great age of hygiene lasted from roughly 1850 to 1950. The front-line troops in the battle for cleanliness were mostly women. Since the 1950s, women have gradually abandoned the home and ventured forth to find external employment. Hygiene standards in the home have inevitably relaxed. Houses are cleaned less often, laundry is done less often, and both are done less thoroughly. Despite the outbreaks in fast-food restaurants that hit the headlines, most foodborne disease actually occurs in the home and goes unreported.

In England, about 1,000 outbreaks of *Salmonella* per year are serious enough to be officially recorded. About 80% to 90% of these involve household exposure and affect only one or two people. The major sources of contamination are sponges and dishrags, which spread bacteria over countertops, dishes, and hands. Bacteria die off in a few hours when surfaces are dry, but they can survive for days or weeks in damp or moist surroundings. Zap your wet sponge or damp dishcloth in the microwave oven for a couple minutes if you want to sterilize it.

About 25% of Americans don't bother to clean their cutting boards after hacking up raw meat or chicken. Vegetables or bread sliced on the same board then pick up bacteria. A few years back, housewives were told that plastic countertops and cutting boards are safer because their smooth surfaces allow them to be wiped clean more thoroughly than wood. No one bothered to do the experiment, however, and the opposite was proven true: Wooden cutting boards are safer than plastic ones. First, plastic might look smooth to you, but on a microscopic scale, it is covered with hills and valleys where bacteria can hide. Second, many wooden boards contain phenolic compounds that are toxic to bacteria. Third, as wooden boards dry in the air, they tend to suck bacteria on their surfaces into pores in the wood. The bacteria might survive, but they become trapped and cannot get back to the surface.

Cannibalism is hazardous to your health

Cannibalism undeniably shortens your life expectancy if you are on the menu. But what if you're the cannibal? Eating people should be highly nutritious—after all, the human body contains all needed nutrients. True enough—if the victim was well fed. But even so, there's a snag. You are more likely to catch some nasty infection from eating another human than from gulping down part of a pig or sheep. Cannibalism is relatively rare among animals, too, and part of the reason is the same; carnivores are more likely to be infected by a disease lurking in a close relative than by a disease that infects another species.

The most spectacular disease spread by cannibalism is kuru, which used to be endemic among the Fore tribe of New Guinea. Kuru is passed on by eating the brains of previous victims. The Fore did not eat their enemies, but they practiced ritual cannibalism of their own deceased kin. Women had the honor of preparing the brains of dead relatives and taking part in their ritual consumption. Consequently, 90% of the victims were women and the younger children who helped them. Kuru develops very slowly, and it can take 10 to 20 years for symptoms to appear. First comes

severe headaches. Difficulty in walking and neural degenera-
tion follow, with death resulting in one to two years. No one
born to the Fore since 1959, when cannibalism stopped, has
developed kuru.

Most diseases result from microorganisms that contain
their own genes as DNA or RNA. However, kuru and related
weird diseases of the nervous system have infectious agents
that contain no DNA or RNA. Instead, infection results from
rogue proteins known as prions. This protein is actually
coded for by a gene belonging to the victim. This gene normally produces a protein that is found on the surface of nerve
cells, especially in the brain. Stanley Prusiner, who won the
1997 Nobel Prize for discovering prions, was ridiculed for a
decade by those who refused to believe that a lone protein
could cause an infection.

The critical property of the prion protein is that it has two
alternative structures. These differ only in the way the protein folds. Occasionally, the properly folded form rearranges
to produce the rogue form of the protein, which somehow
cripples nerve cells and leads to their death. When a rogue
prion protein enters a healthy nerve cell, it binds to its normal relatives and incites them to refold into the incorrect
shape. After most of the prion proteins convert to the bad
form, the nerve cells die. Patches of neighboring cells die,
leaving holes in the brain like those in Swiss cheese. Prion
disease is known as spongiform encephalopathy because the
brain decays into a spongy mass. If the defective prions man-
age to establish themselves, the disease might progress slowly
but is ultimately fatal.

Because the prion is coded by one of the victim's own
genes, it can cause spontaneous, inherited, or infectious disease. There is a very low chance that a prion somewhere in
the brain might spontaneously flip-flop into the bad form.
This happens in about one person in a million. In inherited

prion disease, a defect in the prion gene generates a mutant prion protein that changes into the rogue form more frequently. Infectious prion disease happens when the victim consumes prions that are already in the rogue conformation.

Many details of prion disease are still mysterious. The extremely low frequency of infection and the long period before symptoms appear make study difficult.

Mad cow disease in England

Scrapie of sheep was the first prion disease identified, in 1738. Infected sheep behave strangely and *scrape* themselves against trees and fences, damaging their fleece and skin. Scrapie is transmitted when the remains of dead sheep contaminate pasture. Infection of new victims by prions is difficult, and only certain genetic breeds of sheep are susceptible.

Mad cow disease broke out in England in 1985 and has since killed more than 150,000 cattle. Even more were destroyed to halt its spread. Before 1985, the disease was unknown. Mad cow disease is not "natural," because cows are not naturally carnivorous. However, to avoid waste, animal remains (including the brains) were ground up and added to animal feed—a mind is a terrible thing to waste. Because sheep remains were fed to cows, the emergence of mad cow disease was originally blamed on sheep with scrapie. However, people in England have eaten sheep with scrapie since the 1700s without noticeable ill effects. Nor have other domestic animals caught scrapie, despite sharing the same fields. Experiments confirmed that sheep prions do not infect cows.

It is now thought that a random flip-flop event converted a normal prion into a rogue prion inside a cow's brain somewhere in England in the late 1970s. The rogue cow prions were recycled by feeding animal remains and eventually spread, causing an epidemic. The kuru epidemic in New Guinea presumably occurred the same way. A prion spontaneously misfolded in a member of the tribe a few generations ago, and the disease then spread by cannibalism.

In the last half-century, four outbreaks of prion disease have occurred among farmed mink in Wisconsin. These mink are often fed on "downer cattle," animals that fall down and die out on the range and thus cannot (legally) be fed to humans. Assuming that all four outbreaks were due to cattle with bovine spongiform encephalopathy (BSE, the official name for "mad cow disease"), and knowing the total number of downer cattle fed to mink during this period, probably about 1 in 28,000 had BSE. Taking into account the total number of cows in Wisconsin, this implies that 1 in every 980,000 suffers spontaneous BSE. This is essentially the same rate as for humans with spontaneous prion disease.

If all animals suffer spontaneous prion misfolding at about the same rate, why has mad cow disease not appeared in the United States or other industrial nations that recycle animal remains into feed? One major reason might be that England is much more densely populated, by both people and domestic animals, than the United States. Consequently, animal byproducts were more intensively recycled. Alternatively, the British BSE outbreak might have been triggered by a rare mutation that yielded an unusually infectious and virulent form of BSE.

Why did the epidemic start when it did? During the 1980s, several changes took place in the British rendering industry that processed animal remains. A new American process was introduced that used lower temperatures, and solvent extraction was abandoned. Solvents had been used to remove animal fat as tallow. When new regulations for using flammable solvents were introduced, most rendering plants just gave up solvent extraction rather than buy expensive new machinery. During the 1980s, the addition of meat and bone meal to cattle feed rose from 1% to 12%. This was partly due to the increased cost of imported fishmeal and soybean meal, due, in turn, to a drop in value of the pound.

Experiments suggest that the previous process was not hot enough to inactivate the prions anyway. On the other hand, prions are protected by animal fat, so the combination of high temperature and fat removal might have made the earlier process safer. Perhaps, then, the mad cow epidemic was triggered by the solvent safety legislation. Or maybe it merely resulted from the increased use of rendered products.

The political response

Political responses to health issues can be divided into those that are ideological and those that are nonpartisan and show that our leaders can nobly rise above mere party lines. When a new problem arises in society, especially a novel health issue, the response of the political establishment is pretty much as follows:

1. The problem does not exist.

2. The problem is extremely rare and, in any case, is declining. There are more important things to worry about.

3. The problem has been highly exaggerated by irresponsible activists and popular journalists.

4. There is a serious problem, and we have been doing everything possible to deal with it from the very beginning.

Both the American response to the AIDS crisis and the British response to mad cow disease followed these stages. Consider the British government's statements on what it preferred to refer to as BSE:

> *"Nobody need be worried about BSE in this country or anywhere else."*—John Gummer, U.K. Agriculture Minister, 1990

> *"There is a continued downturn in incidence of BSE."* —Angela Browning, Junior Agriculture Minister, 1995

In 1995, replying to a mother whose daughter had died of BSE, John Major, the prime minister, wrote, "I should make it clear that humans do not get mad cow disease." The deliberate lying finally stopped on March 20, 1996, when Stephen Dorrell, Secretary of Health, admitted in parliament that BSE had infected humans who had eaten beef. By May 1996, BSE had affected 160,000 cattle on more than 30,000 English farms.

After mad cow disease broke out in England, the recycling of animal remains in feed was prohibited and infected herds were destroyed. British beef was banned from many other nations, despite the British government's campaign of secrecy.

Mad cow disease has recently regained attention because a handful of cattle in the United States have been diagnosed positive. Consequently, several countries have instituted restrictions on the import of American beef.

Mad cow disease in humans

When prions are subverted into changing shape, they fold to mimic the prion that triggers the disease. When rogue cow prions infect humans, the misfolded prions are characteristic of mad cow disease. In humans with spontaneous prion disease or kuru, the detailed shape of the misfolded prions is different. This is because the incoming prions form a template upon which healthy prions aggregate after refolding into the rogue conformation.

Calculations based on the history and age distribution of BSE in humans since the outbreak started suggest an average incubation period of about 17 years. The total number of cases expected is estimated to be 200 to 300. These estimates are much lower than many earlier and overly emotional predictions and reflect the extremely low infectivity of prions when crossing the boundary from one species to another.

Despite the hysteria associated with mad cow disease, the number of casualties is very small compared to those who die from more mundane food poisoning from bacteria or viruses. Perhaps the abnormal fear stems from the incurable nature of this disease.

Fungal diseases and death in the countryside

By and large, fungi have gotten a raw deal when it comes to sharing credit for causing disease and altering history. This is for three main reasons. First, fungal infections today generally range from the irritating to the embarrassing, from athlete's foot to vaginal yeast infections. Few people nowadays die of fungal infections. Second, when we think about historical diseases, we tend to focus on spectacular epidemics such as the Black Death, which depopulated Europe in the 1300s, or the smallpox that decimated the Aztecs. Slower-moving diseases attract less interest. Third, in modern societies, little difference remains between town and countryside—everyone has electric lights, flush toilets, and iPods. In addition, we have far more historical records from towns and cities than from rural areas. Consequently, we wrongly assume that medieval peasants died from the same causes as their city-dwelling brothers.

Highly virulent diseases such as plague and smallpox ravaged the crowded city populations of medieval times. Once a vigorous epidemic had started, fleeing townsmen likely carried it into the countryside. Nonetheless, quite a few isolated villages totally missed out on the Black Death, even during the most severe epidemics. Overall, rural areas had less disease, more because of the absence of overcrowding than because of sanitation. Consequently, the population often increased in rural areas even as it declined in the cities.

Although life was healthier in the countryside, we should not be misled by poetic descriptions of pastoral bliss. Although the death rate in rural medieval England was significantly lower than in the towns, it was still atrociously high by modern standards. Life expectancy was around 30 years. A typical pair of medieval peasants might produce a dozen children, yet only three or four survived, hence rural population growth. Medieval peasants endured a life of hard work in unhygienic conditions. They were badly housed and often malnourished. In contrast to city folk, their major exposure to infection was from the microorganisms infecting their livestock and crops. In particular, the major culprits were fungal.

Fungal diseases and cereal crops

Cereal crops such as wheat, barley, rye, and millet are all highly susceptible to diseases caused by microscopic fungi. Such fungi might live as single cells (yeasts and plant rusts) or as thin, multicelled filaments (molds). Most diseases of major crop plants are fungal and are spread by spores carried in air or water. These diseases go by names such as molds, mildews, smuts, rusts, scalds, blotches, blights, and so forth. Although such diseases are unfamiliar nowadays, their short Anglo-Saxon names reveal that these diseases were familiar to ordinary people over many generations.

When grain is harvested and threshed, fungal spores are tossed into the air. When these chores are done by manual labor, the workers breathe in the spores. In the industrial world we are still familiar with black-lung disease of miners. Here, coal dust is breathed in and causes severe irritation and inflammation of the lungs. However, coal particles do not grow and divide. Fungal spores do. If a healthy, well-fed person, who lives in clean, dry, warm housing breathes in a lungful of fungal spores, he might suffer minor irritation, but little else. If you are cold, wet, and poorly fed, the fungal spores will take advantage of your weakness to germinate. They will grow into fungal filaments, which penetrate deep into the lung, destroying human tissue and using the nutrients for their own growth. Once a fungal infection has taken root, there is little chance of dislodging it.

Although fungal infections of the body surface are rarely dangerous, invasive fungal infections are often fatal, even today. *Aspergillus* is a common mold that occasionally invades the tissues of hospitalized patients who are severely weakened by other conditions. More than half of patients whose internal tissues had been invaded by *Aspergillus* died within 3 months, despite treatment with modern antifungal agents.

In medieval times, sooner or later, a poor harvest would come. People would have too little food to last the winter and would be seriously weakened. Add to this a cold, wet winter and inadequate shelter. Any fungal spores waiting in the lungs would see the situation as a heaven-sent opportunity to germinate and get established. So during the Middle

Ages, most agricultural workers (that is, the majority of the adult population) probably eventually died of fungal lung infections.

As technology improved over the centuries, farming became more efficient and agricultural workers came into less intimate contact with spore-bearing grain. Better technology also decreased fungal infection of crops. This both decreased crop losses and reduced the risk of human infection. Farm workers came to live longer and to produce the greater amounts of food needed to feed a larger population. Nowadays, so few humans die of fungal infections that they have almost been forgotten.

Religious mania induced by fungi

Strangely, a few fungi that infect crops also affected religious behaviur in medieval times. This was due not to a massive death toll undermining religious faith, but to a direct effect on the nervous system. The ergot fungus, which grows on cereals, especially rye, produces toxic alkaloids. High doses cause madness or death, but smaller doses cause hallucinations and bizarre behavior.

The ergot alkaloids are a complex mixture of lysergic acid derivatives. The notorious hallucinogen LSD is a semiartificial derivative of lysergic acid. A crude mixture of ergot alkaloids was sometimes used in earlier times to induce childbirth. The effects were erratic because the level of alkaloids in different fungal extracts varied markedly, and this use has been abandoned. However, lysergic acid propanolamide is still sometimes used to control bleeding after childbirth.

The ergot fungus, *Claviceps purpurea*, produces a brown to purplish body that replaces the grain in infected plants. Ergot-infected rye was often made into bread, and outbreaks of poisoning were frequent in medieval times. The severity varied, depending on the amount of contamination and the strain of fungus. Rye was not cultivated on a large scale until around the fifth century, in Eastern Europe and western Russia. From there, rye spread to the rest of Europe. During the Middle Ages, the symptoms of ergot poisoning, but not the cause, became well known. France was especially susceptible because rye was the major crop, and the mild, wet climate helped spread the fungus. In 944 A.D., about 40,000 people died of ergot poisoning in the south of France. Major outbreaks of ergot poisoning then occurred every five to ten years throughout large areas of France and Germany until the 1300s.

Poisoning occurred in two main forms. The most common was called *ignis sacer* (Latin for "holy fire") or St. Anthony's fire, due to burning sensations in the extremities. In severe cases, the constriction of blood vessels caused gangrene of the arms and legs. This resulted in slow and painful loss of the limbs, which in medieval times was usually fatal. The second form caused convulsions and bizarre behavior. Although less common, the convulsive form probably affected society more. It gave rise to outbreaks of dancing mania. The earliest major example occurred in Aix-la-Chapelle, France, in summer 1374, with other episodes occurring over the next half-century.

The authorities responded by playing music because this was believed at the time to have a soothing and curative effect on madness. In this case, the demented victims danced on to the music. Occasionally, the dancers fell out with the religious authorities, and in Liege they ended up cursing the priests. More often the dancers had a variety of hallucinations, which took orthodox religious forms, such as visions of the heavens opening to show Christ and the Virgin Mary

enthroned in splendor. Overall, such visions tended to vali-
date religious belief.

The witch hunts of medieval days might also have been
partly the result of ergot poisoning. In this case, milder and
sporadic poisoning was involved. What evidence is there that
the witch hunts were due to ergot instead of mere super-
stition? First, most witch trials occurred in regions where rye
was the staple cereal crop. Second, the number of witch trials
was higher in years when there was a poor rye crop. When
food was scarce, poor people had little choice but to consume
grain of poor quality, even if it was contaminated with fungus.
Trials were also more frequent in years when the weather
favored the spread of the ergot fungus—that is, cooler and
wetter than normal. By the nineteenth century, major outbreaks of ergot poisoning were largely restricted to Russia.
The death rate in some of these outbreaks was as high as
40%. Ergot poisoning might also have triggered the famous
Salem witch trials of America. The settlers in the Boston area
grew rye under conditions that would have favored ergot.

Although the people of the Middle Ages did not understand the cause of disease, bitter experience had taught them
to regard plagues as contagious. When they encountered illnesses that apparently struck at random, such as epilepsy or
the bizarre behavior from ergot toxicity, they tended to
invoke religious explanations, such as possession by evil spirits or the effect of a witch's curse. The victims of ergot poisoning were often regarded as the victims of witchcraft, not
as witches themselves. Some other unfortunate wretch would
often be blamed—perhaps a physician who was suspiciously
effective in providing relief to the afflicted. Before laughing
too loudly about the credulity of medieval peasants, remember that a substantial proportion of today's Americans believe

that little gray aliens in flying saucers visit Earth on a regular basis. Why is there no tangible evidence of this? Naturally, a conspiracy—not of Satanists, but of government officials!

Over the past few hundred years, rye has been phased out and gradually replaced by wheat or barley. Today rye makes up a relatively minor proportion of the cereals grown. Is the resultant decrease in ergot poisoning partly responsible for the corresponding decline in religious belief? Is it just coincidence that the colder, northwestern parts of Europe, where wheat and barley were preferred over rye, have tended to adopt the plainer Protestantism, whereas both Catholic and Eastern Orthodox Christianity have more decorative images and icons of saints and angels? Such cultural issues are complex, but perhaps biology plays a greater role than previously suspected.

Catastrophes caused by fungi

During the Middle Ages, fungal infections took a steady, continual toll rather than appearing from time to time in virulent epidemics. As with malaria in tropical areas, they rarely caused specific catastrophes, but provided the backdrop to daily life—and death. Nonetheless, fungal diseases have sometimes caused historical catastrophes by destroying crops instead of attacking humans directly.

The best-known case is the Irish Potato Famine of 1845–50, which was caused by the fungus *Phytophthora infestans*, the agent of potato blight. The blight appeared in Ireland in 1845 and destroyed about 40% of the potato crop. The following year, the entire crop was wiped out. In 1845, the population of Ireland was approximately 8.8 million. By 1851, it had dropped by about 30%, due to the loss of 2.5 million people. About half had died; the other half had emigrated.

The potato blight also destroyed most of the potato crops of North America and Western Europe during the same period. The inhabitants of these other regions were less affected because they grew a mixture of crops, including cereals. In Scotland and Scandinavia, the cold winters killed the fungus and the potatoes survived largely unscathed. The Irish suffered most because they had become almost totally dependent on the potato alone.

Rather than describe the history of the potato blight, let us relate this disaster to human disease. The potato blight parallels the effect of virulent plagues on humans. Dense populations of identical or related individuals are susceptible to massive infection and death. As human settlements have grown denser, so have their crops. Worse, many crops are genetically identical, or nearly so, because of centuries of selection for improved properties. This so-called monoculture is highly vulnerable to emerging epidemics.

Human disease follows malnutrition

Another link to human infection is that most of the victims of the potato famine died of disease rather than actual starvation. Lack of food weakened the famine victims, making them easy targets for infection. As the famine proceeded, hordes of starving refugees left their devastated fields for government workhouses. Here they crowded together in conditions that were not so much unhygienic as prehygienic.

The two main killers were typhus fever and cholera. Typhus fever is spread by lice, ticks and fleas. We describe typhus fever at greater length when we discuss its destruction of Napoleon's Grand Army during his failed attempt to invade Russia (see Chapter 6, "Pestilence and Warfare"). The bacteria that cause cholera and dysentery are spread by water contaminated with human waste. These bacteria all cause diarrhea and are spread by it. Even modern refugee camps run by the Red Cross frequently suffer outbreaks of dysentery or cholera. Some 50% to 60% of the deaths among present-day African refugees are the result of such diseases.

Coffee or tea?

Just as potatoes are associated with the Irish, tea is a symbol of British culture. Although tea was important from early colonial times, until the middle 1800s, the British drank more coffee than tea. The switch from coffee to tea was largely due to another fungus, the coffee leaf rust, *Hemileia vastatrix*. Unlike many fungi, coffee leaf rust is highly specialized and infects only coffee. It can be recognized by the yellow-orange powdery spots on the underside of leaves.

Both coffee and its fungal enemy come from North Africa, especially Ethiopia. From there, coffee was taken to many European colonies. In particular, massive jungle clearing for coffee plantations was done in Ceylon (Sri Lanka). The rust soon followed. A British explorer in Kenya first saw coffee leaf rust in 1861. In 1867, the rust hit Ceylon and quickly spread to other countries. The destruction of the major cash crop during the next decade led to economic problems. Several British colonies changed to tea production instead. And so the British switched to drinking tea instead of coffee. During the nineteenth century, British soldiers abroad even made a remedy for intestinal problems by stirring a spoonful of gunpowder into a cup of tea.

For the next 100 years, coffee leaf rust was restricted to Africa and Asia, but South America remained rust free. Then in 1970, the rust appeared in the Western Hemisphere for the first time, in Brazil. Within 20 years, it spread to the rest of South America and the Caribbean. In today's world, coffee beans are the second most valuable international commodity, after oil, and the rust is controlled by fungicides. Curiously, the coffee leaf rust fungus is itself the victim of parasitism by other fungi. These grow on the rust and destroy many of its spores. Whether these "hyperparasites" could be used as biological control agents remains untested.

Opportunistic fungal pathogens

As noted, because fungi cause few deaths in industrial nations today, their role in infectious disease has been neglected. Recently, the infection of AIDS patients by a variety of opportunistic diseases, including fungi, has reawakened interest in fungal disease. Opportunistic diseases result from microorganisms that are harmless to healthy humans but that, as their name indicates, seize the opportunity to kick you when you are down. They include viruses, bacteria, and protozoa, as well as fungi.

Fungal spores rarely penetrate unbroken skin or an undamaged intestinal lining. The vast majority of fungal infections enter via the lungs. To penetrate the finer airways of the lung, spores must be less than about two-millionths of a meter in diameter. Larger particles are trapped and swallowed or coughed out. A wide variety of fungi can cause similar lung infections, although they are rarely able to establish a foothold in the healthy and well fed. These fungi normally live on dead and decaying organic matter. Thus, they rarely possess mechanisms for causing disease in living victims. If by chance they land on organic matter that shows little sign of vigorous life and self-defense, they simply start to feed.

Friend or enemy

The fungi are not our enemies, nor are they our friends. A fungus does what it must to survive. Sometimes this conflicts with mankind's vested interests, but other times the fungi help us, although not intentionally. In the human war against infectious disease, fungi have been one of our most helpful allies. This is because fungi are just as capable of killing bacteria as they are of destroying potatoes or rotting the lungs of a malnourished peasant.

Fungi kill bacteria by secreting chemicals known as antibiotics. The most famous family of antibiotics, the penicillin family, is made by molds. Penicillin is produced by the bread mold, *Penicillium notatum,* after which it was named. Since its discovery in the mid–twentieth century, penicillin has saved many millions of lives, far more than the relatively small number lost by the action of malicious fungi, either by direct attack or via crop loss and ensuing starvation. Unfortunately, a growing problem nowadays is the development of antibiotic resistance among many disease-causing bacteria.

6

Pestilence and warfare

Who kills more?

Although we humans pride ourselves on our ability to destroy each other, most war casualties have been the result of infections, not enemy action. No, not infected wounds, either. Nowadays bullet holes and stab wounds are rarely fatal unless they hit a vital organ. Before antiseptics and antibiotics were available, however, wounds often became infected and frequently resulted in the loss of life or limb. Nonetheless, most casualties before modern times resulted from infections that killed soldiers who had not been wounded in combat—very often soldiers who had not even gotten into combat. Only since the twentieth century have improved hygiene and, to a lesser extent, enhanced firepower allowed humans to kill more people than microbes have.

The first major conflict in which more combatants were killed by other humans than by disease was the Russo-Japanese War of 1904–1906. Even here, this applied only to the cleaner and more hygienic Japanese. The Russians lost more men to disease than to enemy action, and they also lost the war. Whether World War I was a step forward is debatable. The armies on the western front were vaccinated and disinfected, and suffered most of their casualties from enemy action. The eastern front was quite a different story, and infectious disease retained the upper hand. By World War II, mankind had at last progressed to that point at which humans were more of a threat to human life than microorganisms.

Spread of disease by the military

Warfare is generally accompanied with an upsurge in infections, among both the military and surrounding civilian populations. Modern readers probably think the phrase "well-seasoned troops" refers to those with battle experience. Not so. In historical times, smart commanders gathered their recruits together and stayed somewhere safe and well supplied for a few months. During this period, the soldiers shared their infections and became "seasoned." Being ill while properly fed and housed instead of in active service greatly reduced overall casualties from disease. We discussed the principles involved in the spread of disease in Chapter 3, "Transmission, Overcrowding, and Virulence." Three factors are especially important in the spread of disease by warfare: crowding, mixing, and mobility.

- *Crowding:* Armies are crowds of men who eat, sleep, and work in close contact. Such conditions are ideal for the spread of disease. Moreover, hygiene is poor. Soldiers might be crowded together in camps, barracks, or trenches. Sailors are crowded on warships. Troop ships carrying soldiers packed like sardines are even more crowded. Only in recent times have hygienic measures

been adopted. Very early in history, armies were doubtless dirty, but they were also relatively small and so
were not especially prone to disease. As armies grew
over the ages, they provided infectious disease with
ever better conditions for spreading.

- *Mixing:* When armies are assembled, recruits gather
  from separate localities all over the nation or empire.
  As they mix together, they exchange their infections.
  Diseases previously confined to one locality or segment
  of the population are shared. During the American
  Civil War, about a third of the Northern troops fell ill
  before leaving their training camps, mostly due to
  intestinal ailments, including typhoid. In World War I,
  a similar proportion of draftees was rejected as unfit.
  Although the rural population was healthier, rural
  recruits fell prey to infections more often than those
  from towns and cities. The urban population had been
  exposed to diseases such as measles, mumps, chickenpox, diphtheria, scarlet fever, and tuberculosis early in
  life. Many rural folk, especially from remote areas, had
  never been exposed to these diseases. Thus, when
  these populations were mixed, the rural enlistees were
  vastly more susceptible.

- *Mobility:* Armies rarely stay put. Whether they
  advance, retreat, or merely march in circles, they carry
  infections with them. Armies carry disease across land,
  and navies carry disease across the seas. Even when
  armies are stationary, as in siege or trench warfare, new
  recruits are constantly drafted to the front and the sick
  are withdrawn. Prisoners transfer disease from one
  army to another, and deserters carry disease wherever
  they seek shelter. Infections are spread to civilians

along the route of troop movements. In addition, wars often displace large numbers of civilian refugees, who flee to safety either individually or en masse. Ticks, mites, lice, and fleas travel with the humans they inhabit. Disruption caused by warfare also displaces rats and mice, which consequently move around with their accompanying fleas.

Another, though less universal, factor is the chaos generated by warfare, especially among the vanquished. This often leads to food shortages and malnutrition. Crops might be accidentally trampled, seized by the military, or deliberately destroyed. Poor nutrition weakens resistance to infection. Standards of hygiene lapse, especially among armies in retreat and civilian refugees. The chaos factor has probably had more impact in modern times. In medieval times and before, most people were unhygienic and malnourished anyway.

Is it better to besiege or to be besieged?

In the realm of disease, it is undeniably better to give than to receive. Being sealed in a castle or walled city with a lot of other grubby humans, horses, dogs, rats, mice, lice, and fleas was usually worse than attacking. The defenders were obliged to take their livestock inside (together with their diseases) if they wanted milk or fresh meat. The defenders were unable to get rid of human or animal waste unless they were willing to haul it to the top of the walls and pitch it over at the enemy. Sewage and refuse piled up. Dead bodies could not be buried. Rats scavenged in the garbage. Fresh water was scarce and often contaminated.

Although the odds definitely favored the attackers, the downfall of the defenders was by no means inevitable. If a crowded metropolis is besieged by soldiers from rural areas, the urban/rural exposure factor noted earlier might come into play. The city population will have built up resistance to disease, and the rural population will lack previous exposure. A few captives, deserters, or envoys from the city could spread a devastating epidemic among the encircling army.

The siege of Mecca by the Ethiopians (569–570 A.D.) and the siege of Jerusalem by the Assyrians (701 B.C.) are both cases in which the attacker was routed by pestilence and the siege therefore failed. Mohammed was born in A.D. 570, and Islam had not yet been founded. Nonetheless, Mecca was already the holy city of the Arabs, and the Ethiopian Christians were hoping to destroy Allah's sacred shrine, the Ka'aba. An epidemic, probably smallpox, brought them to their knees and killed their leader, Abraha, who rode on a white elephant. In the Koran, we read this:

> *"Have you not considered how Allah dealt with the Army of the Elephant? Did he not foil their stratagem and send against them flocks of birds which pelted them with clay-stones, so that they became like plants cropped by cattle?"*—The Elephant, The Koran

The Islamic expansion of the seventh and eighth centuries spread smallpox throughout the Mediterranean area. Or perhaps we might argue that smallpox cleared the way for Islam to expand, much as it later cleared the way for the European colonization of America. The Islamic Empire crossed from North Africa to conquer Spain in 710 and penetrated into France in 731. By this time, Europe had recovered from the colossal population die-off from the bubonic plague of Justinian's day and was adjusting to smallpox. The Arabs failed to take France but held much of Spain until the next Black Death pandemic, in the mid-1300s. (Even then, the Moors held Granada until 1492, the year Columbus sailed for America.)

Sometimes disease declines to take sides and strikes down besieger and besieged alike. The siege of Kaffa, which resulted in the entry of the Black Death into Europe, is a case in point. The plague spread from the encircling Tartar army to the besieged Italians, demonstrating that epidemic disease is only too likely to travel from one army to another.

Disease promotes imperial expansion

"God is usually on the side of the big squadrons and against the small ones."—Roger de Bussy-Rabutin (1618–1693)

As explained before, civilizations tend to develop resistance to epidemic disease that then acts as a protective shield. Dense populations acquire a repertoire of infectious diseases. The denser the population, the greater its repertoire. Provided that it survives the initial onslaught, its people gradually become resistant to many of these infections. When disease-adapted populations contact other civilizations, the infections are transferred. The result depends on how much previous exposure to these infections the recipients have had. If the recipients are from a sparser culture with no previous exposure, they will be devastated. The most spectacular example is the depopulation of the New World by European diseases. However, similar die-offs on a more restricted scale have probably accompanied the spread of all major civilizations.

Small tribes of early man domesticated dogs to guard against predators and help in hunting. We might view diseases as being domesticated by larger societies. After a dense population center has adapted to a particular infection, this domesticated disease will guard against intruders and help during missions of aggression. When a densely populated city forms, it will have an advantage over smaller neighboring communities. Infectious disease will spread from the big city and decimate the neighboring cultures. Their territory can then be absorbed with less military effort. The invaders will occupy the new territory. If the original inhabitants who remain are too few to regain independence, they will be assimilated and, in time, become citizens of the growing empire. The cycle can now repeat itself. After an imperial nucleus has formed, it tends to grow like a snowball rolling downhill.

The growth of the Roman Empire proceeded in much this manner. Rome absorbed neighboring villages, then small towns, then the other Italic states, and so on. The Romans had the good sense to extend citizenship to the people they assimilated, giving biology a helping hand. Eventually, the expanding Roman sphere came into contact with the other major Mediterranean civilizations and their diseases. Although the halo of urban disease still provided valuable protection against barbarian incursions, it was of little use against other major urban blocks, such as the Greeks or Carthaginians. From this point on, disease no longer favored the Romans. Ultimately, Rome expanded far enough to make contact with a series of novel diseases that had evolved among the dense populations of Asia. The outcome proved disastrous for the Romans. Several major yet still unidentified epidemics, followed by the bubonic plague, depopulated the Roman Empire. Doubtless, the Romans gave the combined disease repertoire of Europe and North Africa to Asia in exchange. This reciprocal export of infection seems to have caused depopulation not only for the Romans, but also for the Chinese Han Empire.

Protozoa help keep Africa black

The most prominent example of disease fighting for the less densely populated side is the special case of Africa. As the original homeland of mankind, Africa retains many of our ancient tropical diseases. Malaria, sleeping sickness, and yellow fever combined to protect sub-Saharan Africa from Arab invaders and then Europeans. The natives of Africa have evolved at least partial resistance to many tropical diseases, and although they still suffer a major disease burden, outsiders are vastly more susceptible.

Tropical diseases have also kept other tropical regions fairly isolated from world culture until recently. What is so special about tropical diseases? Malaria, sleeping sickness, yellow fever, and several less-well-known tropical infections all must be spread by insects that cannot survive the cold. In contrast, diseases such as smallpox, measles, and influenza move directly from person to person. Consequently, they are not confined to particular climatic zones. Eventually, any population that is exposed will develop resistance to these airborne epidemic diseases.

Insect-borne tropical diseases cannot travel outside their own geographical hot zone. The only way people can become habituated to malaria or yellow fever is to migrate into the tropics and settle among the insects that carry these infections. For example, if yellow fever had been able to spread from Africa around Europe and the Middle East, the inhabitants of these colder regions would have built up resistance over several generations. Then when they entered Africa, they would have been resistant because of prior exposure.

Of course, African tropical diseases have spread, but only to the tropical zones of America and Asia. Such transported African diseases have continued to act as a biological force field to protect their carriers. In 1801, the black slaves in the French colony of Haiti revolted. Napoleon sent in experienced French troops who, other things being equal, should have easily suppressed the revolting slaves. Yellow fever, imported from Africa with the slaves, won the war by killing nearly 30,000 French troops. The slaves were mostly resistant to yellow fever because their African ancestors had adapted to it for many generations. Napoleon withdrew, and Haiti has been independent ever since.

Europe eventually conquered Africa because of superior military and medical technology. However, Europeans have never adapted biologically to the diseases of tropical Africa and must constantly take precautions such as antimalarial drugs and mosquito nets to preserve their health. Moreover, the European conquest of Africa has proven transient, with little permanent settlement, except in regions with temperate climates, such as South Africa. The inability of Europeans to live "naturally" in the tropics has hastened imperial withdrawal.

Is bigger really better?

The Persians tried several times to invade Greece. After Darius was defeated at the battle of Marathon in 490 B.C., his son Xerxes returned to Greece ten years later. Xerxes led the

largest army the ancient world had ever seen, supposedly of
800,000 men, although it is hard to believe that these numbers were not wildly exaggerated. Of these, 300,000 rapidly
succumbed to disease. The identity of the disease, or diseases, is not known. Plague and dysentery have both been
suggested. Plague is usually a disease of settled communities,
not of armies on the march. Dysentery seems more likely,
although it mostly causes widespread debilitation with a
sprinkling of fatalities, not death on a massive scale. In any
case, the Athenians annihilated Xerxes's navy—also oversized
and badly organized—at the battle of Salamis in 480 B.C.,
and Xerxes quit his disastrous campaign and went home to
Persia.

Despite Xerxes's example, the idea that the way to win a
war was by assembling the largest possible army of dirty,
badly fed, poorly trained soldiers remained popular among
ambitious leaders. The Crusades of medieval times (from
1096 to the early 1200s) are another example. Religious hate-
monger Pope Urban II started the Crusades, supposedly to
rescue the Holy Land from the Moslems. More likely, the
real motive was to direct attention away from the corruption
and incompetence of Vatican-dominated Europe. High
points include the loss of 5,000 horses (out of 7,000) in an
epidemic while the crusaders were besieging Antioch in
1098. Without their horses, the heavily armored crusader
knights were helpless. Louis VII of France led some 500,000
men on the Second Crusade, and only a few thousand
returned. Richard the Lionheart of England (1189–1199) left
his country to wallow in economic insolvency while he went
on the Third Crusade and failed to take Jerusalem. His massive army was reduced to a handful of survivors by malnutrition and infectious disease. The microorganisms were
definitely on the side of the cleaner Moslems, with their
smaller, better-trained armies.

Disease versus enemy action

Before about 1800, plenty of rough estimates show that infections killed vastly more soldiers than the opposing army. During the 1800s, more detailed records began to appear. In the Crimean War (1854–1856), the British lost ten times as many soldiers from dysentery and typhus as from Russian action. The mere fact that accurate numbers were recorded is an indication that at last the military was beginning to worry about the loss of manpower from disease. The squalor of the Crimean war inspired Florence Nightingale to push for reforms in hygiene, both in the armed forces and back home in England's filthy cities.

By the Boer War (1899–1902), in which the British fought Dutch settlers for control of South Africa, the ratio had sunk to five deaths from disease to one from enemy action. The turning point was the Russo-Japanese War (1904–1906), in which the Japanese lost only one quarter as many men from disease as from enemy action. Over the next decade, the world's advanced countries copied the Japanese: After scrubbing their recruits clean, they inoculated them against typhoid, tetanus, smallpox, and other infections.

In World War I, most armies stayed relatively free of disease. The major exceptions were the Serbs and Russians, whose outdated and ill-disciplined armies suffered massive losses from typhus fever. The only significant epidemic among the Western Allies in World War I was syphilis, and this caused trivial damage in comparison with losses to disease in earlier wars. Venereal diseases are a special case because they are aided in their spread by embarrassment and secrecy. (The Spanish Flu of 1918–1919 killed more humans than the military actions of World War I, yet this was a worldwide pandemic shared by civilian and military alike, not a specifically military problem.) By World War II, antibiotics and insecticides had made their appearance, too. Soldiers and sailors—and sometimes civilians in war zones or refugee camps—were dosed, disinfected, dusted, and deloused. In our modern era, infectious disease kills fewer soldiers than military action.

Since World War II, the major effect of infectious disease on the military of the industrial nations has been indirect. Providing ever more costly medical care to forestall every imaginable ailment has made armies more expensive. The result has been a decline in the advanced nations' desires to intervene in the Third World, especially in the less hygienic parts. Thanks to rising health standards, the citizens of industrial nations have come to expect their full three-score years and ten. American conscripts in Vietnam had little wish to risk losing 50 to 60 years of life expectancy for little convincing purpose.

Typhus, warrior germ of the temperate zone

Which diseases were responsible for slaughtering the armies of Xerxes or the crusaders remains a mystery. As we approach more modern times, the identities of diseases that have overthrown armies start emerging from the mists of history. During the 1300s and 1400s, the Black Death killed soldier and civilian alike.

From 1494 to the early 1500s, syphilis caused major casualties among the military, but typhus soon emerged a far greater problem. In 1494, the French were forced to withdraw from Naples because of an outbreak of syphilis. In 1528, they were still fighting Spain for control of Italy and were forced to retreat from Naples again. This time the French lost 30,000 men to typhus. From the sixteenth century on, typhus became the major microbial player in the military arena.

Note that typhus and typhoid are quite distinct. Historically, they were confused because both cause fever and a rash. An unfortunate consequence of this is the use of two variants of the same original name for the two diseases. However, typhus results from the bacteria known as *Rickettsias*, which can grow only inside animal cells. Typhoid is the result of virulent *Salmonella*, which can be grown in culture. Typhus is transmitted by lice or ticks, whereas typhoid is spread by water or food contaminated with human waste.

Typhus is at home in cooler climates because the lice that spread it thrive in warm, dirty clothing, especially fur or wool. It is difficult today to imagine just how lousy and verminous our predecessors were—even the upper classes. Archbishop Thomas á Becket was assassinated in December 1170, and his body lay overnight in Canterbury cathedral awaiting burial the next day. He had on "a large brown mantle; under it, a white surplice; below that, a lamb's-wool coat; then another woolen coat; and a third woolen coat below this; under this, there was the black, cowled robe of the Benedictine Order; under this, a shirt; and next to the body a curious hair-cloth, covered with linen." As the body grew cold, the inhabitants of these multiple layers began to evacuate. "The vermin boiled over like water in a simmering cauldron, and the onlookers burst into alternate weeping and laughter."

In 1542, 30,000 soldiers fighting for the Holy Roman Empire against the Turks were lost to typhus. In 1556, Maximillian II of Germany took 80,000 men to fight the Sultan of Hungary, but abandoned the campaign after losing most of his army to typhus. Assorted European wars supplied typhus with a steady stream of victims for the next few centuries.

Typhus was carried aboard ships and took a steady, massive toll among the sailors of the competing European imperial powers. It was especially lethal when allied to the poor nutrition and vitamin deficiency typical of long ocean voyages. The British navy stripped, scrubbed, and shaved its sailors before issuing clean clothes. They also issued lime juice to prevent scurvy (vitamin C deficiency). The British consequently lost far fewer sailors than their enemies, especially the French and Spanish.

The most famous outbreak of typhus was the destruction of Napoleon's Grand Army in Russia. We have already noted Napoleon's Haitian fiasco, but this was not his first brush with disease. In 1798, Napoleon invaded and occupied Egypt briefly. More than 40% of his men died, mostly from bubonic plague. Napoleon's army was so weakened that the British easily ejected him from Egypt. However, Napoleon's biggest bacteriological disaster came during his attempt to conquer Russia in 1812. The French Grand Army started with about 450,000 men. By October, only about 80,000 were fit to fight, and Napoleon began his retreat from Moscow. A mere 6,000 men made it back to France. Of course, the cold, poor nutrition, and the Russians didn't help; nonetheless, typhus caused most deaths, with dysentery as runner-up.

After the Napoleonic era, the Western nations got cleaner. During World War I, typhus was confined to the eastern front. The Serbians lost 150,000 to typhus in the first six months of the war, including more than half of their 60,000 Austrian prisoners of war. Paradoxically, this aided the Serbs because the Austrians were so frightened by the typhus epidemic that they stayed out of Serbia for fear of infection. Overall, the Russians suffered most from typhus. During World War I and the Russian revolution that it triggered, the Russians encountered roughly 25 million cases of typhus and suffered 3 million deaths from typhus. (Among all other combatants in World War I, there were five million cases and about half a million deaths.) The typhus outbreak during the

Russian revolution was so severe that Lenin stated, "Either socialism will defeat the louse, or the louse will defeat socialism." In World War II, scattered outbreaks of typhus occurred, especially in concentration camps, but the use of insecticides such as DDT to eliminate the lice that carry typhus virtually eradicated the disease.

Jails, workhouses, and concentration camps

Jails and workhouses resemble refugee camps and concentration camps in many ways. In all of these, helpless humans are herded together, usually under unhygienic conditions. During the eighteenth and nineteenth centuries, typhus spread like wildfire in the overcrowded jails of industrial Europe. Cool, damp conditions, dirty clothes crawling with lice, and malnutrition combined to promote constant epidemics of typhus. Workhouses and orphanages for those without jobs or homes of their own were almost as bad. Ships carrying European emigrants to America or Australia often took aboard more passengers than they were intended for and suffered similar outbreaks of disease. Camps for refugees and prisoners of war were often overcrowded. Any overcrowded institution, whatever its original motivation, is vulnerable to typhus if hygiene lapses. Of course, other diseases also had a field day, but typhus was the biggest killer.

Although overcrowding is the most important factor, such institutions, whether barracks, prisons, nursing homes, or orphanages, have another serious drawback: Their inmates all share resources. They breathe the same air, drink the same water, and eat the same food prepared in the institution kitchens. Consequently, they also share any infectious agents that use air, water, or food as their mechanism for distribution. The steady increase in population, coupled with the tendency to share resources for economic reasons, has made modern society increasingly vulnerable to waterborne and foodborne disease, whether spread naturally or deliberately.

Germ warfare

Burning crops and poisoning the water supply were probably the earliest forms of biological warfare. Tossing dead or rotting animals into wells or waterholes was doubtless reasonably effective. Throughout history, there have been occasional sporadic attempts to deliberately spread infection for military purposes. These have mostly been ineffective or irrelevant. For example, the attempts of white settlers to spread smallpox to American Indians were largely irrelevant because smallpox had already spread by itself.

By medieval times in Europe, cattle infected with anthrax were being hurled over the walls into castles or walled cities to break sieges by spreading disease. Anthrax is a highly infectious cattle disease that is readily transmitted to humans. It causes a high death rate and was probably reasonably effective. Nonetheless, given the state of hygiene in most medieval towns or castles, there was little need to provide outside sources of infection. With plague, typhoid, smallpox, tuberculosis, dysentery, diphtheria, and measles always around, all that was usually necessary was to let nature take its course.

The reason germ warfare has been of little account until recently is that plenty of dangerous infections were already in circulation. If the enemy is crowded and unhygienic, some natural disease will undoubtedly attempt a biological assault without waiting for artificial prompting. Only in our modern disinfected age has deliberately spreading disease become a meaningful threat.

Psychology, cost, and convenience

During the Vietnam War, the Viet Cong guerillas dug camou-flaged pits as booby traps. Within these, they often positioned sharpened bamboo stakes or splinters smeared with human waste. Although it was possible to contract a nasty infection from these, the main purpose was psychological. Worrying about possible booby traps slowed the movements of American troops out of all proportion to actual casualties. Thus, the threat of chemical or biological warfare might have great psychological effect.

Taking protective measures is costly and inconvenient. Vaccinating soldiers against all possible diseases that might be used is expensive and time-consuming. Moreover, vaccines often have side effects. Consider the anthrax vaccine used by the U.S. army. It was approved in 1971, has been thoroughly tested, and is considered relatively safe. It produces swelling and irritation at the injection site in 5% to 8%, and causes severe local reactions in about 1% of those inoculated. Major systemic reactions are "rare," but the vaccine has not been widely used. Vaccination requires six inoculations, plus annual boosters. Although it works against "natural" exposure, it is uncertain whether it would protect against a concentrated aerosol of anthrax spores.

Even without germ warfare, drugs given to troops from temperate countries to ward off malaria and other tropical infections can damage health if taken over a long period. Constant exposure to insecticides can damage the nervous system. Dressing infantry in protective clothing and respirators hampers mobility. In hot climates, extra clothing can also cause heat stress.

Anthrax as a biological weapon

Anthrax is a virulent disease of cattle that infects humans quite easily, causing a high death rate. It is caused by a bacterium, *Bacillus anthracis,* which is easy to culture and forms spores that can survive harsh conditions that would kill most bacteria. The spores can lie dormant in the soil for years before germinating upon contact with a suitable host. In some ways, anthrax is the ideal biological weapon: lethal, highly infectious, and cheap to produce, with spores that store well. As noted, during the Middle Ages, cattle infected with anthrax were sometimes hurled into castles or walled cities to break sieges.

The problem with anthrax is that the spores are so tough and long-lived that getting rid of them after hostilities are over is almost impossible. After your enemy has been eliminated, the idea is to move in and occupy his territory. Unfortunately, anthrax spores persist so long in the soil that they are likely to infect the invaders. Off Scotland is the tiny island of Gruinard, which the British used to test anthrax during World War II. Thousands of sheep were used as victims, and large amounts of anthrax spores were scattered around. Although the island has been fire-bombed and disinfected, it remains uninhabitable even today because anthrax spores still survive in the soil.

Amateurs with biological weapons are rarely effective

In 1995, the American Type Culture Collection (ATCC) shipped a culture of *Yersinia pestis* (bubonic plague) to a member of a group of white supremacists. The activist was convicted of falsifying a federal certification number that he used to fool the ATCC into thinking he was associated with a bona fide institution. In 1998, the FBI arrested the same individual for possessing *Bacillus anthracis* (anthrax). However, it turned out to be a harmless vaccine strain! Ironically, U.S. Army researchers isolated these harmless, nonsporing derivatives of anthrax for use in immunization.

Even more farcical was the report in July 1998 of a plot by Republic of Texas separatists to assassinate President Clinton. Their plan was to use a cactus thorn coated with the AIDS virus, anthrax, and botulism. A modified cigarette lighter would have fired the projectile. This incident illustrates another weakness of germ warfare: the introduction of needless complications. If you are going to shoot someone, why not use an unmodified gun with an ordinary bullet?

The ATCC received a lot of publicity for supplying white supremacists with plague and Iraq with anthrax in the 1990s. Not surprisingly, a variety of proposals for more regulations were put forward. However, any country that possesses hospitals with microbiological laboratories can wage germ warfare. Major hospitals and research centers in all nations possess stocks of lethal microorganisms that are needed for diagnostic comparisons and to prepare vaccines and antisera. This is especially true of poor nations where virulent infections are frequent. Thus, any reasonably informed hospital microbiologist or clinical technician could obtain cultures of dangerous biological agents if desired. This situation makes both export controls and intensified security procedures for research laboratories futile. Indeed, in many third-world countries, the germs of lethal diseases could be directly isolated from infected people or animals during outbreaks.

Which agents are used in germ warfare?

Among the bacterial diseases, anthrax, brucellosis, tularemia, glanders, melioidosis, and bubonic plague have all been considered for use by the military. Assorted viruses have been suggested, including emerging diseases such as Lassa fever and Ebola virus, but the only consistent choice among viruses is smallpox. If we stick to the idea that germ warfare should be cheap and simple, we can eliminate the viruses. Although they seem more intimidating because they cannot be cured by antibiotics, viruses are difficult to culture because they replicate only in cells of other creatures. Although viruses can be cultured in egg yolks or cultured mammal cells, this requires high technology and trained staff. Granted, substantial batches of virus are grown for vaccines in advanced nations, and a relatively small volume of virus preparation could infect millions. Nonetheless, for quick-and-dirty backyard operations, we should stick to bacteria that can grow by themselves in culture.

Brucellosis, caused by *Brucella,* is a disease of cattle, camels, goats, and related animals. The United States developed it as a biological weapon from 1954 to 1969, although its choice seems curious. In humans, brucellosis behaves erratically, and although victims often fall severely ill for several weeks, it is rarely fatal, even if untreated. Tularemia, caused by *Francisella tularensis*, is a disease of rodents, with a death rate of 5% to 10% in humans, if untreated. Melioidosis, caused by *Burkholderia pseudomallei*, is related to glanders (*Burkholderia mallei*), a disease of horses. Melioidosis is a rare disease of rodents from the Far East that is spread by rat fleas. Despite being only *"pseudo"-mallei*, melioidosis is worse than glanders and is fatal around 95% of the time in humans.

During and just after World War II, bubonic plague was popular. Although normally transmitted by fleas, *Yersinia pestis* can be grown easily in culture and can be distributed by spraying. In July 1948, the *Red Star*, the official Soviet Army newspaper, described a captured Japanese germ warfare facility located in Manchuria. It produced nearly a ton of bubonic plague bacteria per month. The Russians claimed that the Japanese had used prisoners for testing, generally with fatal results. The British biological warfare center at Porton Down also kept large-scale bubonic plague cultures on hand for several years after World War II. In the 1960s, the United States might have experimented with spreading plague among rodents in Vietnam, Laos, and Cambodia. As with much information about biowarfare, this is disputed and uncertain. In 1969, President Nixon announced a ban on chemical and biological warfare research. Since then, interest in bubonic plague has faded.

The Soviet germ warfare program is supposed to have concentrated on anthrax and smallpox (including artificial mutants and hybrids). For an industrial nation, preparation of a virus whose particles are fairly stable and long-lived, such as smallpox, is feasible. In addition, the eradication of smallpox led to the abandonment of smallpox vaccination and, hence, the emergence of vulnerable populations. For a third-world nation, a virus would be a dubious choice. Anthrax is a good deterrent, but occupation of territory after anthrax release is hazardous. For the Soviet Union, with its vast expanses of thinly populated land, this consideration was perhaps minimal.

World War I and II

The Germans made some amateurish attempts to infect animals in World War I. German agents inoculated cattle and horses shipped to the Allies from the United States in 1915 with anthrax and glanders, respectively. Similar schemes were tried in France in 1917. German agents based in Zurich were accused of spreading cholera in Italy. No significant effects can be traced to these attempts; whether they were technical failures or the accusations were mere propaganda is debatable.

The Japanese have been accused of spreading bubonic plague in China during World War II. The Japanese probably had the capacity to grow large cultures of plague. There are accounts of attacks by small numbers of Japanese aircraft that dropped cotton rags, rice grains, and other materials supposedly carrying plague bacteria or perhaps plague-infested fleas. Undoubtedly, minor outbreaks of bubonic plague occurred in central China during the war years. Some 100 to 200 cases of plague were reported, mostly fatal. The Chinese government claimed that these were due to biological warfare, and the U.S. Surgeon General at the time apparently accepted this.

The evidence is not convincing. Bubonic plague has been endemic among the rodents of southern China for centuries. The dislocation caused by war easily accounts for sporadic appearances of plague among humans, even in areas distant from the major plague reservoirs. Furthermore, plague bacteria were not isolated from the material dropped by the Japanese, although several attempts were made to do so. In any case, plague is caught either from flea bites or by breathing in airborne bacteria. Most bacteria are killed by stomach acid, so contaminated rice is useless in spreading plague.

Contaminating cotton rags was also pointless. Spraying a culture of plague would be the only efficient way to infect large numbers of people, as the Japanese must surely have known. What the Japanese were doing is puzzling, but it seems unlikely to have been an effective form of germ warfare.

Germ warfare against rabbits

The Australian attempt to destroy rabbits with myxomatosis in the 1950s received much publicity. But rabbits have the distinction of being the first animal targeted for scientific germ warfare by no less a celebrity than Louis Pasteur. This happened in France in 1887. The rabbits had been burrowing above a wine cellar belonging to Madame Pommery of the city of Rheims. Dislodged stones had fallen, smashing bottles of champagne. Pasteur dispatched his assistant with a culture of fowl-cholera. Three days later, 32 dead rabbits were found and the rest had disappeared. Pasteur then sent his assistant to Australia. Here he met with dismal failure. The infection failed to spread among the rabbits. In addition, the cattle breeders were frightened that their herds might be infected and opposed the program. The project was abandoned.

Half a century later, the Australians deployed a virus. The first year, myxomatosis killed more than 99% of the rabbits it infected. Ten years later, it killed only 20%. The tiny proportion of resistant rabbits who survived the first onslaught did what rabbits do best—they produced lots more rabbits. In short, the Australians selected for the evolution of a myxomatosis-resistant rabbit. Today the rabbit population is booming again and the Australians are ready for another round of germ warfare. Rabbit calicivirus emerged in China in 1984 and spread from there to Europe, Africa, and America. It showed its virulence by killing 64 million rabbits on rabbit farms in Italy. The virus spreads among domestic rabbits and then escapes into local populations of wild rabbits. The virus is highly specific and affects only European rabbits (these are often found on other continents, like the rabbits that plague Australia). The fatality rate is about 95%.

Coordinated release of the virus at 200 to 300 sites in Australia was arranged. The Australians hoped this would minimize the emergence of genetically resistant rabbits or of milder virus. Calicivirus release in the drier parts of Australia (Western Australia, Northern Territory, and South Australia) resulted in a 95% death toll within a few weeks, much as expected. In other areas, eradication has been more erratic. The Australians might gain a few years' respite, but the 5% survivors will not take long to rebuild a rabbit population that will be resistant to calicivirus.

Germ warfare is unreliable

Perhaps one reason the major nations so readily agreed to outlaw germ warfare is that it is ineffective. In practice, bullets and bombs are easier to produce and handle than biological weapons. Another issue is that even the fastest diseases, such as pneumonic plague, take at least 24 hours to kill. And 24 hours is plenty of time for a retaliatory nuclear exchange. Another drawback is the problem of delivery. Spraying is the standard method of distributing germs. Unfortunately, this relies on the weather. First, a breeze is needed—and second, the wind must blow in the *right* direction!

During the 1950s, the British government field-tested harmless bacteria. When the wind blew the germs over "healthy" farmland, most airborne bacteria survived and landed alive and well. In contrast, when the bacteria were blown over industrial areas, especially oil refineries, the airborne bacteria were wiped out. Many airborne industrial pollutants are lethal to bacteria and viruses. Even if the wind is favorable, most of the population of an industrial nation is found in cities, protected from airborne germs by air pollution!

Genetic engineering of diseases

Let's take a harmless laboratory bacterium, such as *Escherichia coli,* and make it dangerous. We insert genes for invading human cells. We provide genes for tearing vital supplies of iron away from blood cells. We add genes for potent toxins that kill people in tiny doses. What have we made? An unstoppable disease that will wipe us from the face of the Earth? No, we just converted *Escherichia coli* into its near-relative, *Yersinia pestis,* the agent of bubonic plague. The reason we are not all dying of the Black Death today is not due to any lack of virulence by *Yersinia pestis,* but to modern hygiene. Improving diseases by genetic engineering is of minor significance. What we should really worry about is being in Mother Nature's gun-sights. Any army that neglects hygiene is crying out for disease to thin its ranks. We don't need "new and improved" diseases: Any of the old favorites could handle the mission, given favorable conditions.

7

Venereal disease and sexual behavior

"That depends, my lord, whether I embrace your mistress or your principles."—John Wilkes (1727–1797), replying to Lord Sandwich, who had just told him that he would die either of the pox or on the gallows.

Venereal disease is embarrassing

In recent times, there has been a tendency to replace the ugly term *VD* (venereal disease) with the supposedly less embarrassing *STD* (sexually transmitted disease). Despite this, accurate data on the frequency of venereal disease is hard to obtain, even in advanced nations. The reason is obvious enough: embarrassment.

The history of venereal disease suggests that, over the long term, periods of promiscuity might alternate with periods of puritanism. Diseases transmitted by direct personal contact tend to grow milder over time, as explained in Chapter 3, "Transmission, Overcrowding, and Virulence." Venereal diseases clearly fall into this category, and syphilis is the classic example. When a new venereal disease emerges, there is often an initial highly virulent phase that affects the more promiscuous members of society. Many victims die or are crippled or disfigured. Sterility often results. Babies are born with congenital infections or deformities. Historically, these effects have been typically viewed as punishment for sin.

This, in turn, incites a major response in social behavior, generally couched in religious terms. The religious authorities and moralizers of the day condemn and castigate sexual licentiousness. A variety of restrictions and regulations are often put into place in the name of God or moral purity. Such restrictions themselves probably have little effect, apart from driving forbidden sexual practices underground. During England's Victorian era, respectable middle-class people suffered palpitations at the sight of an exposed female ankle, and some went to the absurdity of covering the exposed legs of wooden furniture with drapery. Yet estimates suggest that as many as 20% to 25% of the female population of Victorian London practiced prostitution at some time during their lives. Admittedly, most took part only for short spells, to tide themselves over during periods of economic hardship. According to W. O' Daniel, *Ins and Outs of London* (1859), based on police and council records, some 55,000 to 80,000 women were engaged in prostitution at any given time. This was approximately 7% of the female population. (The total London population in those days was around two million.) Historical snippets such as this show that the "good old days" when "traditional morality" supposedly prevailed are largely based on superficial appearances.

Eventually, most venereal diseases decline in virulence and, although they may continue to spread, provoke far less fear. Sermons are muted and regulations are relaxed. The last Puritan epoch was the age of syphilis. It began shortly after the discovery of the New World, when syphilis appeared in Europe and lasted until World War II, when penicillin provided a cure. The postwar sexual liberation boom was muted by the appearance of AIDS in the 1980s. This disease is incurable and lethal, and the possibility of an effective vaccine is remote. However, treatment with expensive antiviral drugs can keep AIDS under reasonable control, at least in advanced nations.

This alternation of puritanism with complacency will presumably continue as long as our planet's biodiversity provides a steady supply of novel sexually transmitted diseases. Of course, complications arise. Fluctuations from promiscuity to puritanism and back are characteristic of urban cultures in which sexual morality becomes interwoven with changing fashions in political ideology and religion. Rural cultures tend to fluctuate less in their moral outlook. Economic conditions also have major effects. Thus, the relative area of female body surface exposed to view in Western capitalist society follows the economic cycle. When the economy booms, more skin is exposed; when recessions come, female fashions become less revealing.

Technological advances have also changed sexual behavior. Birth control pills have vastly reduced the risk of pregnancy following sex. The more recent RU486 (the "abortion pill") has reduced the health risk and cost of dealing with unwanted pregnancies. Antibiotics have effectively cured the traditional venereal diseases, syphilis and gonorrhea. All these advances have tended to make sex less risky.

Promiscuity, propaganda, and perception

Have the last century's medical advances really increased the frequency of casual sex? Or by reducing the damage caused, have they just reduced social disapproval? Does puritanism decrease promiscuity or merely drive it underground? Consider that, despite having much higher church attendance and more religious zealots, the United States has a vastly higher incidence of most sexually transmitted diseases than Western Europe.

Recent studies suggest that rates of venereal disease in the United States are up to 50 to 100 times greater than in other industrial nations. Thus, around 150 in 100,000 Americans have gonorrhea, compared to 3 in 100,000 in Sweden and 19 per 100,000 in Canada. Such figures typically generate calls for education and moral leadership. Both gonorrhea and syphilis declined significantly in the United States during the 1990s. Sadly, this trend reversed from 2000 onward, and both infections have been slowly but steadily increasing in incidence.

Is promiscuity really more common nowadays than, say, in Victorian times, or is it merely less stigmatized? In Victorian England, unmarried mothers routinely moved to a new town and explained their lack of a husband by posing as "war widows." As long as infectious disease cooperated in killing off a steady proportion of the troops Britain sent overseas to control its Empire, this disguise was not only credible, but patriotic. Despite the stories our self-righteous age routinely tells of Victorian prudery and the unfeeling treatment of unmarried mothers, those who pursued the patriotic path often received plentiful help and charity. This scheme also allowed unmarried mothers to "remarry" without fuss. We should realize that most intelligent Victorians were well aware of the war-widow ploy. As long as moral convention was not publicly flouted, few cared about unmasking fakes.

The great problem with studying sexual behavior, let alone its more squalid aspects such as the spread of venereal disease, is locating the reality beneath the veneer of social pretense. Both respectability and promiscuity have their pretensions. Men of the world boast of conquests they've never made, while respectable folk deny involvement in whatever current moral conventions disapprove of. Researchers attempting to draw attention (and funding) exaggerate sexual aberrations, and political activists inflate the numbers of homosexuals. Both traditional religious believers and feminists tend to exaggerate the number of child molesters, especially when novel phenomena such as Internet chat rooms are involved. So don't take the information in this chapter overly seriously. Information on venereal disease is much less reliable than it is for less embarrassing diseases.

The arrival of syphilis in Europe

The writings of Hippocrates, a little before 400 B.C., suggest that herpes, chlamydia, and possibly gonorrhea were circulating in Greece and the Middle East at this time. Chlamydia can cause both venereal disease and trachoma, an eye infection often leading to blindness that was especially common in ancient Egypt. Other diseases can have symptoms similar to gonorrhea, so historical descriptions are mostly ambiguous. John of Ardenne gave the first unambiguous descriptions of gonorrhea in England in 1378. Cases were frequent from then on. The ancient world did not have AIDS, nor did it have syphilis, which was brought back from the New World by Columbus and other explorers.

The "typical" cycle for the progress of a venereal disease is largely derived from the history of syphilis. Syphilis almost certainly came from the Americas. Its European debut was among French soldiers assaulting Naples in 1494, just two years after Columbus sailed. The returning French army took it back to France, where it spread rapidly among both rich and poor. The physicians of the day apparently did well out of it. A monk, seeing Thierry de Hery praying on his knees before the statue of Charles VIII of France, remarked that the king was not a saint. "Perhaps not, but I shall never be able to thank him enough for having introduced into France the sickness by which I have made my fortune," replied the doctor. From France, syphilis spread throughout Europe, Asia, and Africa.

Venereal diseases are typically named after a nearby nation that is disliked. Most Europeans called syphilis the "French pox." When syphilis arrived in the Far East it was called *Guangchuang* (Canton ulcers) by people in other parts of China, and "Chinese pox" by the Japanese. The Bible mentions the "Egyptian botch," and although we no longer know what this refers to, the racial slur suggests an ancient venereal disease. This tendency has caused considerable confusion. In the days before syphilis reached Europe, the term "French pox" was often used, at least in England, to refer to gonorrhea.

When syphilis first appeared, it showed two notable characteristics. First, it was highly virulent. Second, its transmission was not limited to sexual means. In some ways, syphilis behaved in Europe like smallpox in America. The American Indians had no prior exposure to smallpox and were highly susceptible. The same was true of Old World inhabitants facing syphilis. Serious symptoms appeared rapidly after infection, and syphilis was often fatal or severely crippling. Note that the whole French army was forced to withdraw from Naples because of so many rapid casualties from syphilis! Over the centuries, syphilis became milder and the disease progressed more slowly. In many patients, it no longer reached its debilitating tertiary stage, and in those unlucky few, it took many years to do so. One of these was America's most famous criminal, Al Capone, who died in jail of syphilis in 1947, aged 48.

For 40 years, from 1932 to 1972 the U.S. Public Health Service carried out a long-term study of syphilis in black men in Tuskegee, Alabama. Although penicillin, which cures syphilis effectively, became available in the 1940s, the roughly 400 patients were denied treatment to observe the long-term effects of syphilis. It wasn't just Hitler's SS who carried out medical experiments on the racially despised. The relative mildness to which syphilis had sunk is illustrated by this trial that lasted for 40 years, longer than the average life expectancy in 1494.

Similar declines into mildness can be seen in other venereal diseases, including gonorrhea and chlamydia. In 1475, Edward IV returned to England after a campaign in France. He noticed the damage caused to his army by gonorrhea. He complained that he "lost many a man that fell to the lust of women and were burned by them, and their penises rotted away and fell off and they died." It is clear from this graphic description that gonorrhea was much nastier then than now. However, even then, gonorrhea was in decline. Although it still picked off promiscuous stragglers, it was no longer capable of determining the overall outcome of a military campaign.

Relation between venereal and skin infections

Syphilis is caused by the spirochete *Treponema pallidum.* Spirochetes are spiral bacteria often found in freshwater or in the intestines or on the skins of assorted animals, including man. Most are harmless. *Treponema* is a skin-dwelling spirochete that has learned how to invade the body and cause disease. Four types of infection are due to *Treponema:* yaws, pinta, bejel, and syphilis. All are passed along by physical contact. Yaws is a tropical skin disease that sometimes penetrates the internal organs, where it causes serious damage that can eventually be fatal. Bejel is a variant adapted to drier conditions and is frequent in the Middle East and North Africa. Pinta is a milder form seen largely in Central and South America.

Precontact infections of *Treponema* in the Americas seem mostly to have been mild and nonvenereal. The form of *Treponema* that came to Europe in the 1490s was probably intermediate between yaws and syphilis. The earliest European cases were probably transmitted as often by nonsexual contact among dirty crowded soldiers as by sexual contact. Yaws has continued to infect skin surfaces where hot, moist conditions prevail. Syphilis adapted to the colder European climate by colonizing a hot, moist region of the body and so became specialized for sexual transmission.

A similar relationship is seen between the skin disease trachoma and genital chlamydial infections, both caused by *Chlamydia trachomatis.* Trachoma is a disease of the body surface, notable for infecting the eyes and causing blindness in severe cases. It is spread by flies, as well as dirty hands. *Chlamydia* is the second most common cause of genital infections today (human papillomavirus, HPV, is the first).

Although complications can occur, genital *Chlamydia* is usually mild and often symptomless. A 1997 survey of 10,000 sexually active American teenagers revealed that 8.6% of the girls and 5.4% of the boys had *Chlamydia*. More recent, but smaller, surveys suggest moderate improvement. Nearly all the males and 75% of the females had no noticeable symptoms. Cold sores on the skin and genital infections caused by Herpes virus are another such pair of diseases caused by variants of the same infectious agent.

Some early Indian texts from 100–200 A.D. mention a "kustha roga" disease that could well be leprosy. Interestingly, they suggest its spread was at least partly sexual. Some medieval European writings refer to lepers as being lecherous and depraved. The historical evidence by itself is unconvincing, but the tendency of skin diseases to generate venereal variants is quite marked. Was there once a venereal version of leprosy?

AIDS is an atypical venereal disease

Unlike the venereal diseases discussed already, AIDS (acquired immune deficiency syndrome) is not derived from a skin infection. HIV (human immunodeficiency virus), the agent that causes AIDS, is bloodborne and is transmitted both by sex and by dirty needles. Partly because of this, in advanced nations, AIDS is confined to a small sector of the population. The AIDS epidemic has probably peaked in the industrial nations but is now the biggest killer among infectious diseases in the Third World, especially in Africa, where it has overtaken both malaria and tuberculosis.

AIDS is not the only bloodborne venereal disease. Several types of viruses cause hepatitis. About one-third of the cases are due to hepatitis B virus, which spreads by the same mechanisms as AIDS and is prevalent among the same high-risk groups. Many cases of hepatitis B are very mild, and natural recovery occurs after a week or two. Other patients suffer fatal liver failure or long-term liver damage. Several other viruses, some only recently observed, behave similarly. Luckily, hepatitis viruses do not damage the immune system or mutate as rapidly as AIDS. For some viral infections, including hepatitis B, effective vaccines are now available.

Another curious example of a venereal disease that arose from a bloodborne disease is a trypanosome disease of horses. *Trypanosoma evansi* is a relative of sleeping sickness that infects horses. Normally, it is spread by bloodsucking flies, but a specialized venereal strain exists in South America that is spread during mating. Entry requires a local break in the skin surface—that is, blood contact. Such minor skin lesions occur quite frequently when horses mate.

Origin of AIDS among African apes and monkeys

The earliest known sample of HIV was found in blood taken in 1959 from an inhabitant of the Belgian Congo (now the Democratic Republic of the Congo). HIV mutates rapidly and at a fairly predictable rate. Comparison of different early strains, including the 1959 sample and another from 1960, suggest that the HIV ancestor appeared around 1890–1920 and had already diverged into different variants well before the AIDS epidemic emerged into public view.

HIV is closely related to SIV (simian immunodeficiency virus), which is found in assorted apes and monkeys. The whole SIV/HIV group of viruses is relatively new. African monkeys living in Africa frequently carry viruses of this family. In contrast, monkeys of African origin living on Caribbean islands do not carry SIV. The ancestors of the Caribbean monkeys were brought from Africa in the 1600s and 1700s, implying that SIV was not in widespread circulation at that time.

Two groups of human immunodeficiency viruses are known. The original, HIV-1, accounts for the vast majority of cases and is the most virulent. HIV-2 is relatively rare and is less virulent. The ancestor of HIV-1 was probably transferred to humans from chimpanzees in the equatorial region of West Africa. Three subtypes of HIV-1 are known at present that probably derive from separate transfer of closely related viruses from chimp to human. HIV-2 seems to have branched off independently from the SIV family and entered the human population later, probably via the sooty mangabey monkey.

Viruses of the SIV family are still being transferred to humans in Africa. Infection usually happens during butchering of chimpanzees and monkeys for meat or, less often, from keeping them as pets. Several cases of newly transferred SIV have been found in recent years by screening blood samples from Africans. Most SIV strains cause little harm either to their original hosts or to humans who acquire them. Likewise, chimpanzees do not suffer from AIDS, even if deliberately infected with HIV-1 from humans. Very rarely, newly transferred SIV strains can mutate rapidly inside the human host, so generating dangerous novel variants. That is, they can become new HIV strains.

Worldwide incidence and spread of AIDS

In the United States, just over a million people are infected with HIV. The epidemic appears to have stabilized, with about 50,000 new infections per year. New AIDS cases mostly result from homosexual transmission, with smaller contributions from intravenous drug use and heterosexual transfer. In the United States, deaths from AIDS dropped from around 50,000 in 1995 to about 16,000 in 2002 as a result of improved therapy.

In rich nations, AIDS is restricted to a small subpopulation and has not entered the mainstream. But in the Third World, AIDS is vastly more widespread. About 60 million people are infected with HIV worldwide, two-thirds of these in Africa. By 1999, there were 2.7 million deaths due to AIDS worldwide, with 2.2 million in Africa. AIDS has now overtaken tuberculosis and malaria as the leading cause of death among infectious diseases, with about three million deaths per year. The real numbers of AIDS victims may be higher. Just as deaths from syphilis were often recorded as "consumption" in Victorian England, in Africa, social stigma has resulted in AIDS deaths often being officially attributed to tuberculosis.

In industrial nations, about 20% of AIDS victims are female, compared to 50% in the Third World. Several factors affect HIV transmission in the Third World that have little effect in advanced nations. One is the effect of other venereal diseases. To cause disease, HIV needs to enter the bloodstream and invade the white blood cells. Crossing the thin intestinal wall is much easier for most viruses than crossing the vaginal lining. Consequently, anal intercourse is much riskier than vaginal intercourse. The risk of heterosexual AIDS transmission is greatly increased if the victim is already infected with another venereal disease that damages the surface of the genital organs. HIV can then reach the bloodstream via breaks in the skin.

Much heterosexual AIDS transmission in Third World nations occurs in people who already have other venereal diseases. In poor nations, these often go untreated. Even in advanced nations, the risk of HIV infection is much greater for those already infected with herpesvirus. Another reason Africa has been hit so hard by AIDS is that infestation by parasitic worms, which is widespread in poor African countries, also makes victims much more susceptible to AIDS.

In poor countries AIDS is also transmitted by contaminated needles. Poor Third World countries cannot afford disposable medical supplies. Reusing needles or other instruments, which are often improperly sterilized, spreads AIDS. It has even been suggested that vaccination programs against other diseases have contributed significantly to spreading AIDS in Africa. Obviously men, women, and children are at equal risk from this mechanism, regardless of sexual behavior. Most such cases are listed as "heterosexual spread" partly through ignorance and partly to avoid responsibility and embarrassment by the local regimes.

Sharing dirty needles also spreads AIDS among intravenous drug addicts in advanced nations. In most European nations, clean sterile needles are available to drug addicts. Consequently, addicts rarely pass AIDS and other blood-borne diseases such as hepatitis to each other. In the United States, antidrug legislation has targeted everything remotely associated with drugs. Although the drug supply has scarcely been affected, such legislation makes obtaining clean needles more difficult. Consequently, American drug addicts more often share used needles and so spread AIDS to a greater extent than in other industrial nations. Although needle-exchange programs now operate in some U.S. states, there are still fewer of them than in Europe.

The Church, morality, and venereal infections

"It is worse to preach immorality than practice it."
—Italo Svevo

Ancient epidemics were often hailed as the judgment of God. Venereal diseases have consistently been regarded as divine displeasure with sexual immorality. In early medieval Europe, the Church theoretically forbade prostitution. In practice, however, prostitution was tolerated. Bathhouses, some run by civic authorities—even the Church itself—operated in many European cities. Many were flimsy covers for prostitution. Medieval cities often restricted prostitution to certain districts and often prescribed special garb for prostitutes, such as yellow scarves or scarlet sashes.

During the early to mid-1500s, bath houses and brothels were shut down all over Europe as the Catholic Church took aggressive measures against immorality. The standard historical explanation is that the Protestant Reformation frightened the papacy into reaction. The Protestants criticized the corruption of the Catholic Church, both financial and moral, as well as its doctrines. The papacy responded by clamping down on permissiveness, in an attempt to assert its moral authority. However, at precisely this time, syphilis exploded across the face of Europe. Much the same happened in China. During the Ming period (1368–1644), prostitution was originally widespread. No lethal venereal diseases were in circulation, and those who indulged in commercial sex washed scrupulously beforehand and afterward. Little social stigma was attached. The arrival of syphilis, beginning in 1505 in the Canton region, changed all this.

When syphilis first struck Europe, its venereal nature was not evident. At first, it was viewed as divine retribution for general immorality, including gambling and neglect of religious duty, as well as sexual license. Only after a century or so did syphilis become a specifically venereal infection. By the eighteenth century, syphilis was known to be sexually transmitted, and contracting syphilis was seen as concrete evidence of infidelity or indulgence in vice. During this period, upper-class London families bribed the medical authorities to list relatives who died of syphilis as tuberculosis victims. Consumption was fashionable, but the French pox was an improper way to die. By the early twentieth century, syphilis had become milder. Other venereal diseases were milder still, and sexual morals were starting to relax. The discovery of antibiotics and contraceptives resulted in further rapid loosening of sexual morals as the risks of both pregnancy and disease faded away. Then in the 1980s, AIDS emerged from Africa.

Moral and religious responses to AIDS

The puritanical reaction to AIDS has been relatively small in the industrial nations, especially when compared to issues such as smoking or abortion. AIDS affects only a small sector of society. Consequently, it is little threat to "respectable" people, as illustrated by Edwina Currie, U.K. Junior Health Minister, who in 1987 remarked, "Good Christian people … will not catch AIDS."

In Africa, AIDS is rampant throughout the whole population and threatens millions of lives. At the 1998 Anglican conference, the African bishops demanded that the church condemn homosexuality in accordance with the Bible. In contrast, most bishops from the United States, where AIDS threatens only a high-risk minority, want to portray themselves as "broad-minded" by condoning homosexuality.

One bizarre result is that several African religious leaders have accused their politically correct Western colleagues of racism, on the grounds that they lack concern for the lives of poor blacks in Africa.

Another consequence has been to promote the spread of Islam at the expense of Christianity. Moslem communities emphasize hygiene, prohibit alcohol (thus avoiding drunken lechery), and have stricter codes of sexual morality. Good Moslem people definitely will not catch AIDS. It is notable that in sub-Saharan Africa, the more northern nations, where Islam is strongest, have significantly lower incidences of HIV than those in the south, where Christianity is predominant.

Nonetheless, Islam is now making major inroads into South Africa. Estimates of religious conversions show that between 1991 and 2004, Islam increased by more than five-fold. The suggested reasons are that Islam provides a response to the AIDS, alcoholism, and violence that are sadly too common in overcrowded black townships. It is in precisely these areas that the highest conversion rates occur.

Public health and AIDS

A common piece of disinformation is that AIDS patients pose no risk to the health of others. Because AIDS is not spread by casual contact, associating with HIV-positive individuals is risk-free. Rather, those who don't have AIDS are a threat to those who do, because they pass on infections dangerous only to those with damaged immune systems. Although true, this is not the whole truth.

Those infected with AIDS are a public health hazard, albeit minor. The risk is not from AIDS itself, but from associated infections. Some of the sporadic cases of tuberculosis in advanced nations trace back to visitors from Third World nations where TB is prevalent. The majority trace back to AIDS patients who have become a reservoir for tuberculosis in many major cities. Tuberculosis is mostly a hazard to children and old people. Allowing HIV-positive individuals to teach in school, under the guise of "human rights," is a possible way to expose children to tuberculosis. Those with damaged immune systems also carry assorted other opportunistic infections, and prolonged treatment has resulted in many of these acquiring antibiotic resistance.

Even today, some religious fundamentalists in America claim that God sent AIDS to punish society for the sin of tolerating homosexuality. Such small-town morality tends to drive homosexuals to the big cities. Conversely, many parents with young children have moved away from the inner cities, into suburbs or small towns. Such people tend to cite violence and drug addiction as the factors influencing their migration, because it is acceptable to object to these in public. Whether explicitly stated or not, such population movements also reduce the exposure of children to the tuberculosis and enteric diseases carried by HIV-positive persons. Whatever the reason, such migration does protect children from infection. Thus, from the 1980s until very recently, life expectancies have risen gradually in New York State (as in most advanced nations), while remaining constant or even falling in New York City.

I am inclined to believe that many religious rituals and behavioral taboos functioned originally as pre-scientific public health. For example, incest, which is forbidden by most religions, increases the proportion of children born with genetic defects. Before the days of modern genetics, such taboos were justified by religious arguments. So why have most societies viewed homosexuality as taboo? Today a large proportion of homosexuals carry enteric diseases, including giardiasis and amebic dysentery, regardless of the presence of HIV. If enteric diseases were more prevalent among homosexuals in historical times also, homosexuality could have been a significant public health risk. Roughly half of infant mortality in modern Third World countries is the result of diarrheal diseases of one kind or another. Societies that tolerated widespread homosexuality might well have had more enteric disease, hence higher infant mortality. Perhaps this is one factor behind the religious taboos most societies have traditionally placed on homosexuality.

Inherited resistance to AIDS

HIV enters human cells in two steps. First, HIV must bind to its receptor, the CD4 protein, which is found on the surface of many cells of the human immune system. Next, the virus unfolds its docking protein to bind a "co-receptor." Only after successfully attaching to the co-receptor can the virus actually enter the target cell. The most important co-receptor is the CCR5 protein.

Although the CCR5 protein is part of the human immune system, it is not essential. About 20% of white people have a chunk of DNA deleted from one of their two copies of the *CCR5* gene. Those with one defective copy of the *CCR5* gene can catch AIDS, but its onset is slowed by several years. About 1% of white people have two defective copies of the *CCR5* gene. These people are resistant to catching AIDS. This mutation is not found in Africans or Asians, suggesting that it is specific to Western Europeans.

Variations in susceptibility to AIDS also derive from other changes in the *CCR5* gene that alter the level of CCR5 protein expressed. An assortment of mutations in other genes also affects the progress of AIDS, although most of these are less important and less well understood.

A rather tragic inverse example is also known, in which resistance to malaria increases susceptibility to AIDS. The Duffy receptor protein is found on the surface of red blood cells. Ninety percent of Africans possess genetic alterations that result in an absence of the Duffy protein. This provides resistance to some versions of malaria. However, it also increases susceptibility to HIV infection by perhaps 40%.

The ancient history of venereal disease

Sin, sickness, and sex are all inextricably intertwined in both the theory and practice of religion. Sin brings down divine punishment, originally most often in the form of sickness. Sex is often viewed as sinful, and certain religious groups regard even legal sexual relations between couples married in church with suspicion. Not surprisingly, religious enthusiasts view promiscuity and illicit sexual practices as bringing down the wrath of the gods and thus as responsible for a particular group of ailments: the venereal diseases. But which came first, the egg or the chicken? Did violation of pre-existing morality spread disease, or are moral codes actually public health measures constructed in response to the spread of venereal disease?

The ancient religions of the Sumerians, Egyptians, Babylonians, Hittites, and others all had a series of male gods with relatively well-defined individual ranks and functions. There were also numerous goddesses, but these were far less individually distinct. Most female divinities had roles in fertility. Sometimes conception, childbirth, animal fertility, crop fertility, rainfall (or flooding of the Nile), and so forth were parceled out to separate goddesses, but often these functions overlapped and merged. Ancient fertility rites often included what today's culture would regard as dissipated orgies of drunkenness and sexual indulgence. Associated with early fertility religions, we find an intriguing custom of sacred prostitution. Women sacred to the fertility goddess made themselves available in exchange for contributions to the goddess's shrine. This was viewed as service to the fertility goddess of their particular culture.

Clearly, certain ancient views of sexual morality were quite different from ours. The current squabbles between those who disagree on issues such as abortion and homosexuality seem trivial compared to the gulf between the ancients and ourselves. Imagine the response from Americans who claim to respect "alternative lifestyles" if a Mormon or Moslem asked for polygamy to be tolerated. The ancients often accepted sexual profligacy without regarding it as an issue in itself. For example, adultery was not seen entirely as a sexual issue. Married women were forbidden sex with other men not merely because adultery was immoral, but because illegitimate children played havoc with codes of inheritance.

The priestly moral code of the Bible comes later than many of the narrative accounts. The law was probably formulated by the priesthood during the later stages of the monarchy. However, many Bible stories show behavior that clearly violates these later, official standards but that was clearly regarded as acceptable by the people of the day. For example, the biblical story of Judah and Tamar includes sacred prostitution. This is viewed not merely as legal, but as morally acceptable by all the characters in this story. Thus, Judah, a respectable and prosperous farmer, openly sends a friend with payment for the sacred prostitute.

Widespread cult prostitution, with little social stigma attached, implies that promiscuity had little medical consequence in those days. In other words, there was low probability of catching venereal disease. Later, as populations got denser and infectious disease spread, the venereal diseases would have been amplified, too.

Populations dense enough to support major epidemic disease originated around 1000–500 B.C. Before 500 B.C., essentially all religions were polytheistic and included mother goddesses and fertility rites. As towns grew and people crowded together, skin diseases doubtless spread. Soon, we may imagine, specialized venereal forms of some skin infections began to emerge. These were spread by fertility rites and sacred prostitution, which became steadily more hazardous.

Recent experience tells us that military expeditions are a major factor in the spread of venereal disease. The imperial maneuvers of the Assyrians, Persians, ancient Greeks, and Romans doubtless helped spread many infections, venereal and otherwise, from 1000 B.C. onward. Increased trade and military campaigns took growing numbers of single males ever farther from their hometowns and villages. Venereal infections that might have burned out in isolated communities could now spread.

The growing hazards of promiscuity led to a steady change in behavior over the next few centuries. Remember that venereal disease frequently causes sterility, even when its victims remain otherwise healthy. Those who clung to the old ways suffered a decline in health and had fewer children. This was reflected in the decline of fertility cults and female divinities. Buddhism began in the crowded cities of ancient India and spread over the densely populated regions of the

Far East. Somewhat later, Christianity spread throughout the Roman sphere of influence.

The collapse of the Earth Mother religions and fertility cults was largely complete before 324 A.D., when Christianity became the official religion of the Roman Empire. By this time, Christianity's main competitor was no longer polytheism or mother goddess worship, but Mithraism, another fairly monotheistic cult. Some 300 years later, the emergence of Islam, the most monotheistic major religion of all, swept away the remaining fertility cults in the less populated parts of the Near East. Today's world-view is basically that of celibate monotheism. The single gods of modern religions are male, like the chief gods of ancient pantheons. But they are no longer even married, let alone polygamous.

8

Religion and tradition: health below or heaven above?

Religion and health care

Religion is often criticized for offering its adherents "pie in the sky when you die." But do religions really hold the loyalty of their followers by offering an afterlife? Few religious believers, however devout, seem overly enthusiastic about moving onward and upward to paradise. I argue that whether a religion is successful in practice depends on protecting its followers from the hazards of life on planet Earth. For most of history, infectious disease was the major killer, so how religion responded to disease was a crucial aspect.

We must distinguish between the supernatural and natural aspects of religion. Most religions offer prayers for the sick. In addition, most religions perform rites designed to prevent infection or cure disease. The intent in either case is to invoke supernatural intervention on behalf of the patient. Religious believers might accept these procedures as supernatural and believe that they cure disease, even if they do not work.

In contrast to these supernatural aspects, religion also provides physical benefits to health, whether consciously intended or not. Religious rites, such as those for segregating the "unclean" or for removing ritual impurity by washing, often promote hygiene. These measures work even though the practitioners might not understand how infections are transmitted and are not even attempting to prevent the spread of disease. Whether dressed in religious trappings or not, medical care, nursing, and hygiene benefit the sick. Survival of the sick is noteworthy, and if any gods were invoked during the health-care process, they receive credit.

Today in advanced nations, religious faith is dwindling. One factor could be that infectious disease is largely a thing of the past, and religion can do little for heart disease and cancer. Miraculous healing, as a formal part of religion, is often mocked nowadays, even by those who attend church. Yet Christian denominations that merely provide moral guidance and bingo sessions are losing their congregations. Despite its primitive aura, physical healing is in many ways the heart—or the lungs and liver—of religion. After all, if an omnipotent god truly exists, miraculous healing should present no problem.

Meanwhile, in many third-world countries, Christianity has displaced traditional religions. When people saw their own culture as inferior to European civilization in terms of life span and infant mortality, many concluded that the Christians' god must be more powerful than their own. They then switched allegiance. Today, Christianity is numerically the largest religion, with Islam second and gradually closing the gap.

Belief and expectation

Why do some disasters shake religious beliefs while others have little impact? Expectation is a major factor. If you mostly live on hot dogs and hamburgers, you view steak or lobster as a treat. But if you live on cornmeal and greens, as many poor Africans do even today, a hamburger is a delicacy. All things are relative. Animals and humans respond not to constant stimulation, but rather to changes, whether in temperature, noise, diet, or prosperity.

If you live in a society where infant mortality is greater than 50% and life expectancy is around 30 years, you are used to losing friends and relatives at an early age. In medieval England, children were not usually officially named until they had survived their first year. Mortality was so high that families avoided getting overly attached to newcomers whose chances of lasting a whole year were less than even. In contrast, in our modern industrial society, with life expectancy at over 70 years, we do not expect children to die before adulthood, and when this does occasionally occur, people become deeply upset.

When a virulent epidemic rages through a population, its effects on religious faith depend on how long people normally expect to live and how frequently they expect to lose friends to infectious disease. Diseases such as malaria that take a steady toll each year in areas where they are endemic do not shock society. Endemic diseases are part of the natural scenery and are viewed much as other constant natural problems, such as the cold of winter or the barrenness of the desert.

In contrast, virulent epidemics resemble earthquakes or hurricanes. An epidemic of bubonic plague gallops through society like the four horsemen of the apocalypse. Within a few weeks, half the population is dead and the corpses are piled too high for the survivors to bury. Society goes into communal shock. Religious belief is threatened. People might lose faith in their gods and look for other, stronger deities. Alternatively, they might blame their priesthood for failing to intercede successfully. A third possibility is that the priests might convince the people that the plague is divine punishment for sins committed by the people or their rulers.

Observations on surviving hunter-gatherers illustrate the link between disease and religious rites. For instance, the majority of Navaho ceremonies concern disease, and a typical Navaho spent 25% to 30% of his waking time on religious activities. Among the San of the Kalahari Desert, the curing dance is the most important ritual. After entering the spirit world during a dance, the shamans plead with their god to cure the sick. In addition, the shamans chase away the evil spirits of the dead, who are believed to shoot the living with invisible arrows of disease.

Roman religion and epidemics

Before the days of the empire, the Romans responded to a series of severe but not devastating epidemics by switching their loyalty from the original Roman gods to Apollo, the Greek god of healing; then Asklepios; and finally Hygieia. Later, when the Roman Empire was devastated by a series of catastrophic plagues, the whole fabric of Greco-Roman polytheism came apart. Thus, between 150 and 500 A.D., the population of Rome fell from roughly one million to a mere 60,000; from 250 to 500 A.D., the population of the empire dropped by half after successive epidemics.

In the early days of the Christian era, there was a massive emphasis on caring for the sick. Undoubtedly, such care greatly reduces the death rate for many diseases such as smallpox, typhoid, and malaria. Although these diseases are life-threatening, the death rate can be as low as 10% among patients who are kept clean, dry, warm, comfortable, and supplied with food and drink. Victims of the same infections who are shunned and left to fend for themselves are much more likely to die. Unlike many others at the time, the early Christians were willing to take the risks of nursing the sick, and their God was credited with great healing powers as a result. The Black Death is a different story. For those infected with bubonic plague, nursing has little effect. Ebolavirus is the same way. Outbreaks of swift, high-mortality diseases such as these proved beyond the control of man or god.

Infectious disease and early religious practices

If we regard disease as the consequence of sin, as many ancient cultures did, we might interpret the Garden of Eden as a memory of the disease-free days before urbanization spread a succession of pestilences. In biblical genealogies, the most ancient ancestors are credited with extremely long life spans. The records of ancient Middle Eastern kingdoms also give exaggerated life spans for the rulers of the earliest dynasties. Is this just hyperbole attached to the memory of heroes, or is it a half-forgotten memory of days gone by when life expectancy was indeed much longer?

In the ancient world, sickness was often thought to result from spirits of some kind invading the body. Some were merely spirits of the dead looking for a new home; others were genuinely evil spirits, or demons, with malevolent intent. Today we tend to think of evil spirits as an archaic explanation for mental disease. This could be because, in our own culture, explaining brain malfunction has lagged behind understanding other, more physical illnesses. But to those who lived long ago, swellings, spots, and rashes were just as mysterious as mental aberrations. We should also remember that many infectious diseases produce fever and delirium if left untreated, thus obscuring the gulf between physical and mental conditions.

As late as the nineteenth century, diseases such as typhoid, diphtheria, and scarlet fever were blamed on invisible vapors from smelly drains and other sources of putrefaction. It was not just the poor whose dwellings stank:

> "I have met just as strong a stream of sewer air coming up the back staircase of a grand London house from the sink, as I have ever met at Scutari...."—Florence Nightingale, Notes on Nursing, 1859

Blaming evil-smelling vapors is not so different from invoking evil spirits. Both are nebulous and invisible. And today, instead of invisible spirits, we have microscopic germs, carried through the air or water and still invisible to the naked eye.

Worms and serpents

The earliest form of infectious disease whose cause primitive people could actually see was infestation by parasitic worms. Consequently, worms that "poisoned" people from the inside, and snakes that poisoned people by biting, were regarded as relatives; the same terms were often used for both. Moreover, serpents were sometimes associated with the devil and sometimes with medicine. This derives from the ancient viewpoint that pestilence, or perhaps just a single disease, belonged to a particular god.

It was not so much that disease was evil and healing was good. Rather, sending plagues and removing them both came under the authority of the same god, who used disease as a means of retribution for human disobedience. Although most deities could intervene if cajoled with prayers and sacrifice, it made most sense to beg the god who sent disease to remove it. Thus, in many ancient cultures, the god of pestilence is also the god of healing. Later, a division of labor generated deities concerned solely with the healing aspect of disease. For example, Apollo, who both dispensed and withdrew pestilence, was succeeded by Asklepios, who was concerned only with the healing aspect of disease.

Sumerians, Egyptians, and ancient Greece

Religion and medicine were not distinct in early civilizations, and treating disease involved religious incantations as often as procedures that we regard today as medical. The Sumerians, who founded urban Middle Eastern civilization around 4,000 B.C., generally considered sickness due to three main kinds of spirits. These were the ghosts of the dead, genuine demons, and spirits that were the hybrid offspring of demons and humans. Individual disease was wrought by evil spirits wandering around on their own, whereas large-scale plagues were sent by angered gods.

Although any god could send disease, the specialist was Nergal, the Sumerian god of pestilence. Nergal represented the blazing sun at noon, as did his Egyptian opposite number, Sekhmet, the lioness goddess, who breathed fire at the Pharaoh's enemies. She was also called Lady of Pestilence.

The valleys of the Tigris and Euphrates and the marshlands
of the Egyptian delta were fertile for both agriculture and
disease. Nergal was kept so busy that he needed 14 assistants
(more properly, 7 pairs) to give mankind fever. Sekhmet, too,
is linked with the holy number seven. The seven arrows of
Sekhmet brought evil fates, especially as disease.

The ancient Egyptians had several solar deities. In addition to the supreme sun god, Re, several goddesses represented aspects of the sun. Bastet, the cat goddess, was the
friendly warmth of the sun, whereas Sekhmet was the noon-
time sun. Hathor, the great mother goddess, usually depicted
as a cow, also represented the all-seeing sun, or "eye of Re."
As such, she was sent by Re to punish mankind. As Re got
old, he started to worry that mankind was plotting against
him. So Hathor went down to Earth and started to slaughter
the conspirators. She drank their blood, which transformed
her into Sekhmet. Naturally, sun gods and goddesses work
only during the day, so Sekhmet went back to heaven for a
good night's rest. Meanwhile, Re and the other gods had got
cold feet. If mankind was wiped out, who would provide the
gods with sacrifices? So Re told his priests to put red dye into
7,000 pitchers of beer. They poured it out in the desert, and
when Sekhmet came down the next morning, she lapped up
the red beer, thinking it was blood. In her drunken stupor,
Sekhmet forgot about exterminating the rest of the human
race and reverted to the kinder and gentler Hathor.

At their New Year's festival, held just before the flooding
of the Nile in July, the ancient Egyptians attempt to appease
Sekhmet and ward off her "seven arrows." In addition to the
main sacrifices are many offerings of alcoholic drinks. The
drunken feast that follows supposedly re-enacts Sekhmet's
distraction by using red-dyed beer. The timing is appropriate.
Although the flooding of the Nile is vital to agriculture, it also

spreads epidemics. For several weeks, slow-moving water covers everything. Bacteria and viruses that cause dysentery and diarrhea are spread by contaminated water. Malaria-carrying mosquitoes breed in stationary or slow-moving water, and water snails spread the parasitic worms of schistosomiasis (bilharzia).

In ancient Greece, Apollo was both sun god and archer god. As with Sekhmet, his arrows represented pestilence, and he was invoked for healing together with his son Asklepios and granddaughter Hygieia. The Greeks felt it was unmanly for Apollo to shoot women, so they became the first to introduce affirmative action into medicine. Artemis, the twin sister of Apollo, carried a bow and arrows, which she used for bringing disease upon women. The general idea of demons spreading disease by shooting invisible arrows lasted through the centuries. In medieval England, sick animals were referred to as "elf-shot" and were treated by holy water and singing masses.

Hygiene and religious purity

One practical advantage of attributing sickness to spirit possession was that the sick were seen as unclean from a religious viewpoint. In the biblical book of Leviticus, religious impurity itself is viewed as at least partially contagious. Thus, menstruating women, who were ritually impure, could spread their impurity to other persons they contacted. Here again, we see the overlap between a "spiritual" condition and infection. Those regarded as unclean were often quarantined or excluded from society until their symptoms (or their souls) departed. Though often cruel, quarantine greatly reduced the spread of contagious disease. Because most people in antiquity died from infections, quarantine was beneficial overall.

The Sumerians regarded the spirits of the dead as home-less rather than malicious. Sickness then resulted from the misfortune of providing refuge for a homeless spirit rather than deliberate sin. The more rigid Semitic cultures that fol-lowed were less tolerant of the sick. Disease was seen as divine punishment, and sickness and sin became intertan-gled. Sometimes the individual was guilty; sometimes the guilty party was the parents or the tribe or nation as a whole. Somewhere there was sin, and, increasingly, the priest's job was to uncover it and pin the blame on someone.

A similar change from a freewheeling outlook to a nar-rower attitude occurred during the growth of the Christian church. Most early church fathers believed that demons assaulted even innocent Christians because the demons were evil. According to St. Augustine (354–430), "All diseases of Christians are to be ascribed to these demons; chiefly do they torment fresh-baptized Christians, yea, even the guiltless, newborn infants." Thus, caring for the sick, by either secular or spiritual means, opposed the demons and was therefore commendable.

Later the church saw disease more as punishment from God. This at times led to the extreme position, rarely stated explicitly, that curing disease by nonreligious means was a blasphemous attempt to usurp God's authority and, there-fore, was itself sinful. This attitude sometimes had disastrous effects. Even as late as 1885, during a smallpox epidemic in Montreal, the Catholic Church opposed vaccination. The Abbé Filiatrault expressed the official view, "If we are afflicted with smallpox it is because we had a carnival last winter, feasting the flesh…it is to punish our pride that God has sent us smallpox." Eventually, the rising death toll prompted a reversal of the official position.

Financial motivations sometimes lurked behind these attitudes. For example, in 1547, Pope Leo X sold tokens bearing a cross and the inscription, "He who kisses it is preserved for seven days from falling sickness, apoplexy, and sudden death." To be fair, most secular remedies of the day left much to be desired. The physician of the Habsburg Emperor Rudolf II (1552–1612) provided this recipe against plague: "Desiccated toads and pulverized chickens. The menstrual blood of a young maiden. White arsenic, pearls, and emeralds from the Orient. This concoction is to be baked into a toad cake and then worn next to the heart in an amulet."

The Protestant Reformation greatly improved matters by returning to the position of the early Church Fathers and emphasizing that disease came from Satan rather than being approved by God. When faced with the opinion that taking medicine was sinful, Luther asked, "Do you eat meat when you are hungry? Even so you may use physic, which is God's gift...."

Protecting the living from the dead

Disease presented our ancestors with a ticklish technical problem: What to do with the corpses? The earliest hunter-gatherers constantly moved around. Whether they buried their dead within their territory, left their remains in a sacred cave, or merely abandoned them is lost in the mists of time. Later, when agriculture spread, and humans settled in one place, the problem became more acute. Corpses are a source of contagion. The earliest communities began to bury their dead. In the earliest cities, the dead were often buried beneath the floors of inhabited dwellings.

Çatalhöyük (now in central Turkey) was the world's first real city. Nine thousand years ago, it boasted several thousand occupants living in several hundred mud-brick buildings. It lasted for slightly more than 2,000 years and disappeared before the Bronze Age. It was apparently the center of local Neolithic culture, although its political relationships, language, and religion are still obscure. The remains of more than 60 people have been found in burial pits beneath the floors of the living quarters in Çatalhöyük. The bodies were tightly folded with the knees close to the chest, and the pits were plastered over. Some pits were reopened later to accept more bodies. Although this was better than leaving bodies exposed, burying them so close to where people lived and prepared food was not ideal. However, many infectious diseases emerged only after agricultural civilizations created dense populations. When Çatalhöyük first arose, few present-day epidemic diseases had yet appeared, so burying the dead under the floor was still relatively safe.

As urbanization spread, infectious disease increased in frequency and corpses became steadily more of a liability to public health. In response, the dead were buried a safe distance from the living, and graveyards outside the town walls became standard. Later, cremation came in vogue in some societies. The ashes were sometimes kept in urns, sometimes buried, and sometimes scattered to the four winds. Whether burnt or buried, the dead were no longer a focus of infection. These activities were all dressed in religious trappings and given spiritual explanations. Members of those primitive societies often feared that the spirits of the recently deceased would come back to harm them. To avoid this, the dead were properly buried or cremated. Although our culture now thinks of ghosts having clanking chains and haunting castles, the original problem was that corpses were a public health hazard unless properly disposed of. Over the ages, ever more complex ceremonies have accreted around procedures such as burial, but avoiding the spread of disease has remained a critical issue.

In many early societies, contact with dead bodies caused ritual uncleanness. Becoming ritually clean involved purification by washing, often followed by a period of quarantine. The relevance to public health is obvious. Thus, many religious rites, especially those for ritual purity, had positive effects on hygiene.

Diverting evil spirits into animals

A student of mine once claimed that the more people you give your cold to, the faster you recover yourself! Although this was meant as a joke, primitive cultures doubtless noticed that as one person recovered, others fell sick. One obvious interpretation was that evil spirits were moving from one person to another. So what could be more sensible than to break the chain of human infection by diverting the evil spirit into an animal. Even today, the people of the Ewe tribe of West Africa transfer disease-causing spirits to chickens and then chase away the chickens with brooms.

In the Bible, we read of the scapegoat ceremony. A goat, chosen by lot, was designated for Azazel, a demon who lived in the wilderness. The chief priest laid his hands on the goat's head and transferred the sins of the people of Israel into the goat. Such a transfer made the scapegoat unclean, so it could not be sacrificed to God. It was driven away into the desert. It is particularly noteworthy that the priest was instructed to remove his vestments and bathe his body immediately after driving away the scapegoat (Leviticus 16:20–25). Note that this use of animals to remove disease was not a form of sacrifice. The animals that received the disease-causing spirit were never killed, as their role was to carry the evil spirit far away.

Other Middle Eastern cultures had similar rites to transfer sin or impurity into an animal, which was then banished. In the ancient world, where disease was regarded as a manifestation of divine displeasure, infection and iniquity were inextricably bound together. Although the biblical scapegoat carried "transgressions and iniquities," we should remember that punishment for sin often came as pestilence.

Ancient Hittite texts prescribe such rituals "if pestilence afflicts the army or the land of Hatti." One such Hittite ritual used animals of two species, a bull and a ewe—the first in case a male deity had sent the plague, the second in case the deity was female. The scape-bull and scape-ewe were marked with colored wool and then, very sensibly, driven into enemy territory. Do we see here the beginnings of biological warfare?

The Hittites were Indo-European intruders into the Middle East. Did they borrow the idea of a scape-animal from the Semitic peoples of the Middle East, or did they import it? It is interesting that the ancient Vedic culture of India, also Indo-European in origin, practiced similar rites. Demons blamed for disease were transferred to animals, not during official communal ceremonies, as with the Hittites and Israelites, but on an individual basis using small animals that ordinary people could afford or catch. For example, the demon Takman, which caused assorted fevers, was often banished into a frog, which hopped away:

> *"Homage be to the deliriously hot, the shaking, exciting, impetuous Takman! Homage to the cold Takman, to him that in the past fulfilled desires! May the Takman that*

returns on the morrow, he that returns on two days, the impious one, pass into this frog!"
—Atharvaveda VII: 116

In the *Odyssey*, when Homer refers to the long illness of Odysseus, he uses the phrase "wasting long away." The term for "wasting away" is *tekomenos,* a word remarkably similar to the Takman of the Vedas. Although historians have sometimes guessed that "tekomenos" refers to malaria or tuberculosis, there is no way to be certain.

Cheaper rituals for the poor

Few people in ancient times could afford to eat meat except on special occasions. Nor could most victims of disease afford an expensive animal for a healing ritual. So the Assyrians, and probably most other Middle Eastern cultures, had a downmarket, vegetarian, version of the procedure. In the "surpu" ritual, the priest smears the patient with flour. The flour is then wiped off and thrown into the fire. The idea is that the flour absorbs the impurity (or evil spirit), which is then destroyed by the fire. Finally, the priest sprinkles the patient with water.

In China during the Ming period (1368–1644), fever was attributed to malevolent spirits, and a similar but more vulgar rite was performed to expel them. The patient was given a pill of cinnabar plus the seeds of plants containing emetics, held together by bee's wax. A fire was lit and cinders placed all around the patient. When the emetic started to work, the patient was supposed to vomit into the fire. As the fire consumed the vomit, the priest made a magical sign over the fire and the evil spirits were killed.

Rather more radical was the practice of cutting a hole in the skull. This procedure, referred to as trepanation, was practiced as early as the Bronze Age. It started as a ritual performed on corpses, presumably to help speed the spirits of the dead on their way to the afterlife. Then it was used on the living to allow evil spirits to exit. Remarkably, many of the victims survived, to die later of something else, as indicated by the signs of healing around the holes in their skulls. By the time of Hippocrates (around 400 B.C.), trepanation was used to treat bruising or fracture of the skull, mostly due to being bashed on the head during war. As late as the nineteenth century, certain primitive tribes still trepanned cases of convulsions or chronic headache to allow evil spirits to escape. Be grateful for aspirin!

Vampires, werewolves, and garlic

Among the "evil spirits" seen on TV today are werewolves and vampire bats. It is widely known, at least to those who imbibe their medical information from the silver screen, that garlic protects against these terrors of the dark side. Sadly, however, the characters in modern-day horror movies have forgotten how to use their garlic correctly. Onions and garlic, especially, contain allicin and related smelly sulfur compounds. Despite the risk of bad breath, garlic should be eaten, not festooned in bunches around the heroine's bedroom. Allicin is a potent antimicrobial agent; in particular, it cures amebic dysentery. In ancient times, garlic was widely used to drive away the "evil spirits" that caused intestinal disturbances. This is remembered in distorted form in folk tales and the modern Hollywood legend.

In eighth-century Japan, eating large amounts of onions and related vegetables to combat diarrhea was not merely a folk remedy, but was recommended in official state directives. In the year 737 A.D., Japan suffered a great smallpox epidemic. Among the directives issued by Ki no Ason, Great Liaison of the Right, senior 4th rank, lower grade, was the following: "If diarrhea should develop, boil onions and scallions well and eat many." (Note that the diarrhea was due to secondary intestinal infections of those weakened by smallpox.) Allicin works well against intestinal infections because it is poorly absorbed and much of it remains in the intestines.

Divine retribution versus individual justice

Who should be punished for a sin or a crime? Should only the criminal be punished, or should his family be included? Ancient writings often give the impression of great injustice, in the sense that innocent people were frequently punished along with the guilty. Many law codes of the ancient world reveal an apparent contradiction. Most punishments carried out by human authorities were inflicted on the guilty individual alone. Thus, Deuteronomy 24:16 commands, "The fathers shall not be put to death for the children, nor shall the children be put to death for the fathers; every man shall be put to death for his own sin." In contrast, when the gods themselves punish someone, both in the Bible and in other ancient cultures, they often strike down the criminal, his family, servants, and even domestic animals.

This paradox is readily understood once we remember that infectious disease was regarded as the means by which the gods usually punished mankind. It was a matter of simple observation that infections appeared mysteriously and generally affected several people in close contact. If the gods struck someone with a virulent infection, chances were good that his wife and children would catch it, too. In the Bible, divine punishments are often reserved for those whose sins were secret and would have gone unpunished if God had not seen and acted. The book of Leviticus calls for those whose impurity goes unnoticed by the priesthood to voluntarily give sacrifices. The implication is that sins can be hidden from man, but if God intervenes, the punishment will be worse. This was a strong inducement to wrongdoers to confess, even when crimes were unsolved. Unfortunately, this way of thinking led to the belief that those who were struck down by infections must have committed hidden sins. Worse, they must also have declined the opportunity to confess and make voluntary restitution by sacrifice.

Although monotheism is usually regarded as a step upward from polytheism, from a medical viewpoint, it was a step backward. The idea that assorted evil spirits inflicted infections comes closer to the germ theory of disease than later rationalizations. Under monotheism, the victims of disease were thought guilty of secret sins, despite lack of evidence. In contrast, polytheism often regarded the sick as unlucky victims of some passing demon rather than as evildoers. Consequently, treatment of the sick was more humane. In practice, the common people have tended to retain a belief in spirits and demons even in officially monotheistic societies. Is having a demon driven out any less rational than being assumed guilty of invisible sin? During the European Middle Ages, when the educated establishment regarded disease as the result of an imbalance among the four humors (blood, phlegm, bile, and black bile), the common herd stuck to a "primitive" belief in contagion.

The Middle Ages reveal another uneasy compromise between official monotheism and popular polytheism, namely witchcraft. When humans or cattle miscarried or produced offspring with genetic defects, and when infectious disease or food poisoning struck, especially if the symptoms were unfamiliar, witchcraft was often blamed. Blaming such misfortunes on witchcraft allowed the victims to be declared innocent of sin while retaining belief in a single all-seeing God. Although this Jesuitical maneuver relieved the victims of infection and poverty from being blamed for their own misfortunes, it also implied that somewhere there lurked witches who needed to be hunted down. A new form of scapegoat was thus born.

The rise of Christianity

As discussed earlier in this book, around 500 B.C., the first major population centers became large enough to keep virulent epidemic diseases in circulation. Several major population crashes largely lost to history probably occurred. During and following this general period, major religious changes happened. Local city-state and fertility cults were replaced with religions shared by both aristocracy and commoner and both urban and rural populations. I have already argued in Chapter 7, "Venereal Disease and Sexual Behavior," that the emergence of sexually transmitted infections played a major part in the transition away from fertility cults.

The decline of classical civilization and its replacement by Christianity was one of the greatest cultural changes in European history. The two main aspects were the collapse of the Roman Empire and then the loss of faith in traditional polytheistic religion. Christianity grew up during a period when overcrowding was followed by pestilence. The success of Roman culture resulted in territorial expansion and population growth. Population density and the ease of communications, especially via the famous Roman roads, increased markedly. Both factors prompted the growth and spread of infectious disease.

The debilitation of large sectors of the population by endemic malaria greatly weakened the Roman Empire. Several massive outbreaks of pestilence followed, culminating in bubonic plague, which exacerbated the situation. These diseases affected both agricultural production and the availability of recruits for the military, leading to political decline. The failure of classical religion to cure the sick or halt the spread of pestilence caused many to lose faith in the gods of traditional religion. Why did Christianity emerge triumphant from this historical hot zone? Why not polytheism, with a new family of healing deities? Why not a mother goddess cult? Why not Mithraism?

We have relatively little information about history's losers, and much of this is biased. Presumably each of these other religions failed in some way, although we can only guess how.

In his *The History of the Decline and Fall of the Roman Empire,* Edward Gibbon cites the zeal of the early Christians, their doctrine of future life, and their pure and austere morals, among other reasons. More recently, Daniel Reff in *Plagues, Priests and Demons* (2005) has compared the demographic collapse at the time when Christianity was adopted in Europe with the population crash on the American continent triggered by the arrival of the Spanish. He regards epidemics as the "most powerful evangelisers of all."

To its credit, Christianity did tackle the issue of disease. The Gospels place heavy emphasis on healing the sick, and early Christians had a reputation for caring for those who were ill. Miraculous healing was an integral part of early Christianity, not merely in theory, but in practice. Even if we assume that miracles are impossible, can we still explain the success of Christianity when faced with ancient epidemics? I think so.

First, let's remember that no one expected early Christians to heal broken legs or cure tumors. Second, early Christians sincerely believed that they would go to heaven and that life on Earth was merely a temporary interlude. Nonetheless, they also believed in helping the sick and, due to their belief in an afterlife, were willing to take risks that nonbelievers avoided.

I believe that the critical point lies in the Christian custom of caring for the sick. Even severe infectious diseases are rarely anywhere near 100% fatal. Thus, diseases such as smallpox and typhoid can kill anywhere from 10% to 50% of their victims. The death rate depends partly on the strain of microorganism causing each particular epidemic. Nevertheless, the actual survival rate is also greatly affected by whether the afflicted are abandoned to their fate or cared for. Members of a Christian community who fell ill with virulent infections were much more likely to receive care than the rest of the population. Consequently, their probability of surviving was much greater. Furthermore, for those unfortunates who did lose their families to disease, the church provided a new family.

By the fourth century A.D., Christianity was spreading vigorously. It was especially popular among women, often those from the upper classes. Fabiola, a widow from a distinguished Roman family, founded the world's first free public hospital in Ostia, the port city that serves Rome. Fabiola originally went to the Holy Land, to join the Christian scholar Jerome who was translating the Bible into Latin with the help of several rich female religious activists. However, the Huns burst into the Middle East, and Jerome and his clique fled to Rome. After returning, Fabiola organized other rich women into founding a hospital. They not only contributed money, but also took part themselves in doctoring and nursing. Fabiola personally collected poor patients off the streets. This was no task for the squeamish. According to Jerome, "They have leprous arms, swollen bellies, shrunken thighs, dropsical legs…their flesh gnawed and rotten and squirming with little worms…."

Fabiola died in 399 A.D. and was later made a saint. When you consider the positive contributions and personal bravery of these Christian women, it is not surprising that early Christianity gained respect. Both the increased survival of the ordinary Christians and the sincerity of those who died while nursing others must have greatly raised Christianity in the public esteem. Few of us really want to die, even those who believe in an afterlife. Whether the cures were miraculous was not the point—they were attributed to Christianity.

Coptic Christianity and malaria

The Copts were a Christian sect centered in North Africa. They wrote their earliest religious texts in Coptic, a language derived from ancient Egyptian. Later, the Copts also wrote in Greek. Although modern-day Christianity dislikes admitting it, these early Christians practiced what can only be described as a form of magic. Spells were written on papyrus sheets, which were then folded into long strips and worn as amulets. Many of these were designed to protect their wearers against disease, and they invoke not only Jesus Christ, but also a mixture of saints and often also Jewish demons.

Here is part of a fifth-century spell, originally written in Greek on an amulet (Oxyrhynchus 1151) and designed to protect a woman named Joannia from fever (undoubtedly, malaria):

Flee, hateful spirit! Christ pursues you; the son of God and the Holy Spirit have overtaken you. O God of the sheep-pool, deliver from all evil your handmaid Joannia, whom Anastasia, also called Euphemia, bore. ... O Lord, Christ, son and Word of the living god, who heals every disease and every infirmity, also heal and watch over your handmaid Joannia, whom Anastasia, also called Euphemia, bore, and chase away and banish from her every fever and every sort of chill—quotidian, tertian, quartan—and every evil.

Similar spells were found down to the eleventh century, often containing magical formulas from the medieval Jewish cabala, mixed with more orthodox Christian terminology. These spells illustrate the great importance of both malaria and magic among the Christians of North Africa. They also confirm that early Christianity was in many ways a healing cult.

Messianic Taoism during the collapse of Han China

During the early centuries of the Christian era, when successive epidemics weakened the Roman Empire, a similar process occurred at the far end of the Eurasian landmass. Early Chinese civilization was centered in the temperate north, especially in the valleys of the Yellow and Yei rivers. (Southern China, though watered by the more famous Yangtse Kiang River, was civilized much later, largely due to tropical diseases.) From 170 A.D. onward, massive plagues preceded by floods hit the Yellow River valley of northern China. Depopulation was followed by peasant revolts and political turmoil, resulting in the collapse of the Han Empire around 220 A.D.

Just as Christianity emerged from the chaos in the Roman world, a messianic religion based on Taoism emerged in the disintegrating Han territories. By 184 A.D., the Taoist sect of the "Great Peace" had 360,000 armed followers. Chang Chiao, who claimed healing powers, led them. These Taoists worshipped the lord Huang Lao, a hybrid of the fabled Yellow Emperor, Huang Ti, and a deified Lao Tzu, founder of the original Taoist movement. They believed that disease was the consequence of sin, and they distributed healing amulets at the spring and autumn equinoxes. Unlike Christianity, this messianic version of Taoism failed to survive over the long term. Buddhism, from India, displaced it over the next few centuries.

Buddhism and smallpox in first-millennium Japan

During the years 735–737 A.D., a massive smallpox epidemic swept through Japan. The smallpox epidemic was preceded by a famine in 732–733, and the resistance of the population was doubtless lowered. Although sparsely populated regions were scarcely affected, the death rate was 70% or greater in some of the most crowded areas. The overall mortality for the whole of Japan was probably at least 30%. Smallpox was on the rampage in China during the fourth century or earlier and had moved to Korea by the mid-500s. It presumably moved from Korea to Japan: It was first reported in the port of Dazaifu, which is on the Japanese coast opposite the Korean peninsula.

The depopulation of Japan resulted in several major reforms in the areas of taxation, farming loans, and land tenure. It also had a great effect on religion. Emperor Shomu had been brought up as a Confucian. At the beginning of his reign, Buddhism, an import from India via the Chinese mainland, was tolerated but strictly controlled. When the famine struck, Shomu felt responsible: "The rivers are dry and the five grains have been damaged. This situation has come about because of our lack of virtue."

When the plague of smallpox followed, Shomu was even more certain he was to blame: "Recently untoward events have occurred one after the other. Bad omens are still to be seen. I fear the responsibility is all mine." Shomu responded to the crisis by donating massive sums and ordering Buddhist temples to be built all over Japan. Although the economy was already tottering as a result of so many taxpayers dying in the epidemic, his daughter, Empress Shotoku, followed his example. Like her Roman counterpart, Fabiola, she cared deeply for the sick. Sadly, her excessive contributions helped bankrupt the state. Despite these unfortunate financial side effects, Buddhism was instituted at the expense of Confucianism.

The European Middle Ages and the Black Death

The Black Death is especially terrifying. Unlike most infectious diseases, nursing has little effect on the fatality rate from plague. Until antibiotics became available, the death rates for the bubonic (60%–70%) and pneumonic (99%) forms of plague remained unchanged by any treatment. As noted earlier, in the major Roman epidemics of the second and third centuries, the Christians made major progress because nursing greatly reduced the fatality rate. However, neither nursing nor prayer stopped the relentless march of the Black Death.

The effects of the Black Death on religion were complex and, in some ways, contradictory. The inability of the Church to stop the plague or cure the sick resulted in a great loss of faith, not so much in God as in the religious establishment. To be fair, although many of the higher clergy fled, the high death rates among the ordinary priests indicate that most of them performed their duties until the end. Many writers of the day, including William Langland (1322–1400), noted the unworthiness of the higher clergy. As a supporter of a purified Christianity, Langland saw their worldliness as a threat to the Holy Church and remarked, "So we need an antidote strong enough to reform these prelates … who are hindered by their possessions." Education in the 1300s was largely under ecclesiastical control, and doctors of medicine were therefore taught and licensed by the Church authorities. Thus, the failure of medicine to cure the plague was also associated with the Church.

Unlike in the Roman era, no viable alternative religion was waiting for an opportunity to take over. The result was fragmentation of authority within the realm of Christianity rather than the infiltration of a new religion. One aspect of this was a great upsurge in the veneration of previously obscure saints who were supposed to have healed plague victims. Shrines were richly endowed, and new religious brotherhoods formed themselves around these healing saints. In a way, this was a reversion to the polytheism of the Romans and Greeks, when many gods and goddesses hawked their wares in a marketplace of theologies. But instead of autonomous gods and goddesses, a multiplicity of saints remained within the bounds of Christianity. Many ordinary people believed that God was displeased, presumably because the established church or Christian society was impure and corrupt in some manner. Outbreaks of violence occurred, directed at Jews, loose women, lepers, and other outcast groups. Sometimes these groups were accused of actually spreading the plague; at other times, they were accosted for polluting society by their very existence, so bringing down the wrath of God.

In the long term, the greatest effects on religion were indirect and took place over the next two centuries. The Black Death shook the feudal system apart and freed up Western society. Lack of manpower led to mechanization and a readier acceptance of new inventions, such as printing. The collapsing feudal system spurred the growth of nationalism. This, in turn, led to local rulers resisting the centralized control of the papacy. Hence, the religious reforms of Luther in the early 1500s found widespread support, especially among the northern countries. The prime example of this is, of course, Henry the Eighth of England, who split away from the Catholic Church and founded the Church of England. When the monopoly of the Vatican was broken, the growth of religious freedom was free to proceed. Slowly, our modern forms of parliamentary democracy and industrialization emerged.

The Great Plague of London

The Great Plague of 1665 was the last time an epidemic of bubonic plague ravaged London. Although it was nowhere near as terrible as that of the 1300s, it was still terrifying. The government instituted public prayer and days for fasting and public confession of sins. The churches were crowded with people imploring God to stop the pestilence. By the 1600s, British Christians were split between those loyal to the Church of England and the Dissenters (Puritans, Presbyterians, and other nonconformist sects). During the crisis, they prayed together in each other's churches. When the plague passed, the barriers among the separate denominations gradually arose again.

During the early days of the outbreak, London was rife with fortune tellers and soothsayers who, for a small consideration, would tell you your chances of surviving the plague. These prophets for profit merged imperceptibly with conjurors who claimed to possess magical cures and a multitude of quacks selling secret and infallible remedies of a more "scientific" nature. Many Londoners took to wearing charms and amulets to ward off the evil spirits causing the pestilence. These amulets were remarkably similar to those the Coptic Christians used in earlier centuries. They included magic words, such as *Abracadabra,* signs of the zodiac, and the Jesuits' mark of *IHS* (*Iesus Hominorum Salvator,* Latin for "Jesus, Savior of Men").

Loss of Christian faith in industrial Europe

It is a common misconception, especially in the United States, that intellectuals led the movement away from religion. If anything, the reverse was true. Just as political correctness finds its most avid supporters on today's university campuses, the intellectuals of earlier times generally went along with the religious establishment. He who pays the piper calls the tune. By the time Darwin proposed the theory of evolution, organized religion was already losing its grip. In England, the world's first industrial democracy, the first demographic group to desert religion en masse was the urban working class. As industrialization proceeded, there was a bulk flow of population from the countryside into the manufacturing towns. These migrants mostly came from rural areas where religion was a significant part of community life. As they settled in the towns, they tended to leave religion behind.

Villages have a sense of community, and middle-class suburbia values respectability. Both are major contributors to religious conformity. But other factors were at work. The overcrowded urban poor were the most susceptible to infection. Thus, those who were most often the victims of plague and pestilence abandoned a religion that seemed increasingly ineffective. Rural populations were more spread out, and the rural poor often escaped the worst effects of epidemics circulating in the towns. The more prosperous lived more spaciously and more hygienically, so they, too, paid a lower toll to infectious disease. Villagers who prayed together in their parish church for the plague to pass them by often had their prayers answered. In contrast, the inner-city poor who attended church were as likely to be infected there as anywhere else. The great epidemics of the age of industrialization were aided and abetted by the crowding from increasing population and urbanization. Thus, the urban working classes were alienated from traditional religion.

Cleanliness is next to godliness

> *"We should hear no longer of 'Mysterious Dispensations,' and of 'Plague and Pestilence' being 'in God's hands,' when, so far as we know, he has put them into our own."*—Florence Nightingale, *Notes on Nursing,* 1859

During the nineteenth and twentieth centuries, technology took the field against infectious disease. Clean water, sewers, flush toilets, toilet paper, soap, antiseptics, warm and dry housing, and better nutrition all combined to reduce the spread of infectious disease. Toward the end of the 19th century, Louis Pasteur declared, "It is now in the power of man to cause all parasitic diseases to disappear from the world."

Although this was a trifle optimistic, the gains in life expectancy and general health were impressive. Already reeling from the onslaught of civil engineering, infectious disease took another massive beating from the successive discoveries of vaccination in the late nineteenth century and of antibiotics in the mid–twentieth century. Life expectancy in the advanced nations today is nearly twice that at the start of the nineteenth century.

Not only did religion fail to cope with epidemic disease, but science was successful where religion failed. Religion lost its monopoly on healing, and medicine became an independent, secularized profession. Before the modern era, organized religion jealously guarded its right to dispense healing and exert control over life and death. Today even those who still practice religion routinely go to a doctor when they are sick. We sometimes hear the saying "doctors acting like God." While intended to puncture pomposity, there is a deeper truth here. As long as scientific medicine is effective, many people today feel little need for supernatural intervention.

9

Manpower and slavery

Legacy of the last Ice Age

During the last Ice Age, the Bering Strait between Asia and North America became a land bridge joining the two continents. Primitive Asian tribes wandered across into North America and migrated southward. As Earth warmed up again, the ice retreated, and some 10,000 years ago, contact between America and Asia was sundered. For some ten millennia, the inhabitants of the American continent remained isolated from the rest of mankind. During this critical period, most of the epidemic diseases characteristic of the Old World made their appearance.

About 500 years ago, contact resumed when Portugal and Spain, soon followed by the other European naval powers, discovered and conquered the Americas. The result was one of the most spectacular population crashes in history. During the century following contact, the indigenous population of the American continent declined by some 95% due to epidemics. Although the European invaders inflicted some military casualties, the overwhelming number of fatalities were due to infectious disease. The indigenous population of America had no previous exposure to the diseases circulating among Europeans and dropped like flies. Historians ask whether the European conquest would have been successful without the help of infectious disease. Did the Europeans deliberately spread epidemics? What role did religion play in the response to the imported epidemics? In contrast, biologists wonder why no American diseases were capable of wiping out 95% of the population of Europe or Asia. For that matter, what diseases did precontact Americans suffer from?

The New World before contact

From a biological perspective, two outstanding issues are noteworthy. First is the apparent lack of major epidemic infections circulating in the Americas before contact. Second, although the American civilizations were technologically still in the Stone Age, they produced remarkably dense populations, especially in Central America. Some accounts suggest that the Aztecs and Incas had almost no major epidemic diseases and lived unusually long lives before European contact. Other sources recount major epidemics, often following periods of famine.

Archeological data indicates that life expectancies in pre-Columbian America were short, perhaps around 24 years in the early classic Maya period and declining thereafter. These estimates are even lower than for medieval Europe and raise the question of what pre-Columbian Americans died of. The dense agricultural populations of Mesoamerica depended excessively on maize (corn) as a staple foodstuff and suffered from dietary deficiency due to a shortage of meat and fresh vegetables. Consequently, their susceptibility to a variety of intestinal and respiratory diseases was greatly elevated. Overcrowding and poor sanitation also contributed.

Several serious insect-borne infections were present in pre-contact America: Chagas disease, Leishmaniasis, Carrion's disease, Lyme disease, and Rocky Mountain spotted fever. These are all caused by protozoa or bacteria and are restricted in range by the insects that carry them. What the Americas lacked was any major virus disease that spread from person to person. Though hard to believe, this appears to be confirmed by the fact that no epidemics from America devastated Europe. The only major disease the Americans contributed seems to be syphilis. Although syphilis is a serious problem, it does not cause virulent, fast-moving epidemics such as smallpox, measles, or influenza. Tuberculosis and perhaps typhus were also present on the American continent before Columbus made contact.

Little information seems to be available on the precontact rate of infant mortality or its causes. Several respiratory and gastrointestinal infections that can kill infants but have only mild effects on adults probably were in circulation in America and were carried back to join the large number of similar infections already circulating in Europe without attracting much attention.

Indigenous American infections

Examination of pre-Columbian corpses from tropical areas of the Americas has shown moderate levels of infestation with parasitic worms and protozoa. However, inhabitants of tropical regions of Africa or Asia typically have much heavier parasite burdens than precontact Americans. In particular, the three most fearsome tropical diseases, malaria, sleeping sickness, and yellow fever, were absent from the Americas.

Chagas disease is found only in South America and is caused by the protozoan *Trypanosoma cruzi,* a relative of the trypanosome that causes African sleeping sickness. Chagas disease causes fever, chills, muscle pain, and nosebleeds. Many recover after this initial phase, but other victims develop a chronic disease of the internal organs, especially the digestive tract and heart. Today about 17 million Latin Americans are infected, and approximately 50,000 die each year. DNA sequences specific for *Trypanosoma cruzi* have been identified in mummified corpses that are several thousand years old. The mummies are from the Atacama Desert, found where Peru and Chile meet. This area has served as a burial ground for about 9,000 years for Chilean and Peruvian Indians, who buried their dead in shallow graves. The Atacama Desert is so dry that space scientists use it to simulate Mars. Corpses buried there have been desiccated and preserved, although they were not deliberately mummified. About 25% of the mummies tested positive for Chagas disease, compared to 10% to 15% of the modern-day population of Chile.

Chagas disease is spread by the kissing bug, which lives in cracks in house walls. This inch-long vampire emerges to suck blood from its human victims at night. The trypanosome is spread when the bug's droppings infect the wounds it made to suck blood. Charles Darwin probably contracted Chagas disease during his famous voyage around South America on the Beagle. Consequently, he was a semi-invalid later in life. This could be the major reason he settled down to write *The Origin of Species* instead of gallivanting off around the world on more naturalistic expeditions.

Another insect-borne South American infection is Carrion's disease, also known as Oroya fever. Carried by sandflies, it is caused by *Bartonella,* a member of the degenerate rickettsia group of bacteria (other species of *Bartonella* cause "trench fever" and "cat scratch disease"). The death rate is about 40% nowadays. Some of the survivors suffer continuing skin lesions, referred to as *verruga Peruana*. Carrion's disease is solely a human disease. This contrasts with another indigenous American infection, Lyme disease. This is spread by ticks that normally live by sucking the blood of wild deer. Lyme disease is primarily an animal disease and infects humans only by accident. When humans penetrate woodland areas inhabited by deer, ticks carrying Lyme disease sometimes bite them. Although Lyme disease is debilitating, it is rarely fatal. Currently, Lyme disease is largely located in the northeastern United States, but it is spreading slowly. In pre-Columbian times, sporadic infections with Lyme disease presumably occurred but were of little overall importance.

Lack of domesticated animals in America

Pre-Columbian American populations were substantial. Perhaps as many as 20 million to 30 million inhabitants were present in what is now Mexico, 10 million to 20 million in the Inca Empire, and another 10 million to 20 million in North America. Archeological evidence and skeletal samples suggest that these high populations were relatively recent and were severely straining the natural resources available to societies with stone-age technology. As large animals grew scarcer, the people became more dependent on maize. As in the Old World, sedentary agriculture supported a denser population at the cost of a lower-quality diet, with less meat and more cereals.

When Old World populations crowded together in this manner some thousands of years earlier, a series of epidemic diseases emerged. From an Old World perspective, the New

World was overdue for such epidemics. Some evidence suggests that epidemic diseases were beginning to appear. The Aztecs recorded several outbreaks. In particular, a drought followed by famine and disease occurred in the early 1450s, fewer than 50 years before the Europeans arrived. The descriptions are too vague to identify any disease with certainty; however, typhus acting on a weakened population is a plausible suggestion.

Virulent epidemic diseases cannot emerge from a vacuum. Indeed, most Old World epidemic diseases originated from infections of domestic animals. The immigrants who populated America split off from the rest of mankind before most animals were domesticated and had only the domestic dog to take with them. Very few animals were domesticated after entering the American continent. These were the turkey, Muscovy duck, guinea pig, and llama and alpaca (both from the same original wild progenitor). These animals did not exist in huge herds before domestication, and their numbers remained relatively small afterward. Llamas, for example, never spread beyond the Andes. Large herds of horses and cattle did roam the New World until around 11,000 years ago, when the last Ice Age ended. Whether these herds were exterminated by early Indian hunters or by the changing climate is debatable. In either case, their demise meant a lack of animals suitable for domestication and, hence, the absence from the New World of mankind's major source of novel infectious diseases.

The first epidemic in the Caribbean

Although smallpox normally takes pride of place in the litany of pestilence that struck down the indigenous peoples of the

Americas, it was not actually first ashore. In 1494, an epidemic spread from Columbus's ships and ravaged the island of Hispaniola (today split between Haiti and the Dominican Republic). From there, it was spread to Cuba, Jamaica, and other Caribbean Islands partly by the Spanish and partly by islanders fleeing from Hispaniola. About a third of the Spaniards fell ill, though few fatally. Vast numbers of the islanders died. The identity of this disease has remained puzzling to this day. Malaria, smallpox, yellow fever, and bacterial dysentery have all been blamed, yet none fits the bill. These diseases did assault the Americas in due course and were recognized when they arrived. Several investigators have suggested an intestinal infection of some sort, based on the symptoms observed.

A recent theory blames influenza carried by pigs taken aboard Columbus's fleet in the Canary Islands. The extremely rapid spread and the high proportion of people infected are typical of flu. In addition, most of the symptoms would fit with a virulent outbreak of swine flu. However, there are also arguments against this. An influenza epidemic did invade the New World in the 1550s and was unambiguously identified. Furthermore, influenza in the 1500s was still new enough among Europeans that the death rate was around 20%, far higher than seen in 1494 among the Spanish. It is hard to imagine a virus as incredibly transmissible as flu failing to spread to the mainland. Yet the 1494 epidemic does not seem to have moved beyond the Caribbean. It is also hard to imagine that a full third of the Spaniards remained uninfected by the flu until they made landfall on Hispaniola. We would expect most of the Spanish to have been infected and developed immunity during the early part of the voyage, with one or two stragglers barely keeping the virus in circulation until they reached America. Although an aberrant strain of influenza remains a reasonable contender, the identity of the 1494 outbreak is still uncertain.

Epidemics sweep the American mainland

The first major epidemic to harry the mainland started when smallpox reached Hispaniola in 1518. From there it was a short step to Mexico, where smallpox arrived in 1520, just in time to save Cortez from an Aztec counterattack. The Aztecs, with overwhelming numbers on their side, had driven Cortez out of their capital city, Tenochtitlan, and things looked bleak for the conquistadors. Smallpox arrived with the Spanish relief expedition, and disease fought alongside the Spaniards. The Aztecs were devastated. Those whom smallpox did not kill were immobilized by shock. From then on, the legions of Old World viruses raced ahead of the conquistadors. By the time Pizarro reached the Inca Empire in present-day Peru, smallpox had already done its work, arriving in 1525–1526. The ruling emperor and his immediate heir had both succumbed to smallpox, and civil strife over the succession to the Inca throne had ensued. Unlike Cortez, Pizarro met no significant military resistance.

Rough estimates suggest that around a third of the total population died of smallpox. The death rate was doubtless higher among those crowded in large cities, whereas many smaller, isolated communities escaped the worst effects. This is remarkably similar to the first arrival of smallpox in Japan in the 700s. In the cities, two-thirds or more died, and the overall death toll was about one-third. Thus, there was nothing magical about the effects of smallpox on the Amerindian population. Exposure of Old World populations to new and virulent infections has had much the same effect.

Measles followed smallpox, and the contrast between European and American susceptibility was even greater. Smallpox killed Europeans, albeit less often than Amerindians, but measles was rarely lethal to those of European descent. The measles epidemic of 1530–1531 raged through the Aztec domains and then followed smallpox into the Inca territory in South America. In 1546, a third epidemic followed whose identity is still uncertain. In any case, eventually 95% or more of the indigenous population of the Americas was exterminated by these successive epidemics.

Interestingly, the next major epidemic was rather different. Influenza ravaged Europe in 1556–1560. An estimated 20% of the population died in England, and fatalities were probably comparable in the rest of Europe, although records are less complete. Yes, 20%. Influenza was still a relatively new disease to Europeans at that time and was still dangerous. An outbreak of "coughing violence" was also recorded in Japan in 1556, and large numbers reportedly died. Then as now, influenza presumably came from China, where different virus strains from pigs, poultry, and people hybridized generating new variants. The flu epidemic of 1556 reached America in 1558 and so was the first true worldwide pandemic to achieve a major death toll in both the Old World and the New World.

A variety of other Old World diseases, such as mumps and diphtheria, migrated to the New World over the next couple centuries. Smallpox and measles, the two biggest killers, broke out every so often. Relatively isolated tribes often survived unexposed for several centuries. The Mandan tribe of North America was reduced from several thousand to only 30–40 in 1837. They were besieged by the Sioux and crowded together, unable to get away, when an epidemic broke out. The Cayapo tribe of South America was safely

isolated until a single missionary visited them in 1903. By
1927, the tribe had shrunk from about 7,000 to less than 30.

The religious implications

The colossal death toll among the indigenous inhabitants
convinced the Christian invaders of North America that their
occupation of the Americas was approved and foreordained
by God. Conversely, the Aztecs and Incas felt that their own
gods were angry and had disowned them in favor of the
Europeans. Neither side understood the nature of disease,
but the fact that the same diseases that decimated the American Indians caused so few casualties among the Europeans
was decisive to both. The Aztecs and Incas were demoral-
ized, and the survivors were easy prey to Catholic missionar-
ies spreading Christianity. It made sense to worship the God
of the victors.

The Puritans who founded the Plymouth colony believed
that God cleared away the Indians for their benefit. Only
divine intervention could account for the incredible mortality
among the Indians while leaving the colonists essentially
unscathed. According to Puritan leader Cotton Mather, "the
woods were almost cleared of these pernicious creatures, to
make room for a better growth." Perhaps it is hardly surprising that the Puritans regarded themselves as favored by God.
In the days before science revealed microorganisms to
human view, there was no other convincing explanation.

One outstanding theological problem at that time was the
origin of the indigenous Americans. If Adam and Eve had
been created in the Garden of Eden, somewhere in the Middle East, how did America get its pre-Columbian population?
Many Puritans believed that the devil had lured the Indians

to America. Here in splendid isolation, the devil was free to rule over the Indians without worrying that the Gospel might intrude.

When smallpox, measles, and other virulent viral diseases first appeared, they doubtless devastated Old World civilizations, much as America was devastated in the years following Columbus. The responses to the pestilence that swept the American continent yield some insight into the behavior of ancient Europeans and Asians, who faced similar catastrophes long before accurate records were kept. Both invaders and the invaded agreed that the massive death toll could have had only one cause: God's will. The invaders were convinced that the Almighty had foreordained their occupation of America. When we find historical Old World cultures utterly convinced that God is on their side, we might well wonder if it is for the same reason. For instance, the Bible tells how the Angel of the Lord killed 185,000 Assyrians in one night, so relieving the siege of Jerusalem by Sennacharib the Assyrian.

Deliberate use of germ warfare

In 1763, during the French and Indian War, Lord Jeffrey Amherst ordered blankets contaminated with smallpox to be distributed among enemy tribes of Indians. Captain Ecuyer gave blankets from the smallpox hospital at Fort Pitt to two Indian chiefs. Although there was a severe outbreak of smallpox among the Indian tribes of the Ohio Valley, it is debatable whether the blankets caused it. Smallpox was already in the area, hence the local smallpox hospital from which the blankets came. Many other stories exist of items contaminated with smallpox or measles being given to North American Indians, and tales of settlers deliberately passing on these diseases are part of colonial folklore.

Until the late nineteenth century, disease was thought to originate in dirt, sewage, refuse, and swamps, and the odors and vapors arising from them. Thus, in 1699, a German missionary commented, "The Indians die so easily that the bare look and smell of a Spaniard causes them to give up the ghost." Many colonists may well have believed that contact with something soiled, such as blankets in which a smallpox victim had slept, was necessary to spread the disease. Having seen the devastating effects of smallpox or measles on Indians, they naturally concluded that the Indians had somehow obtained contaminated materials. So many of the tales are probably later attempts at rationalization more than historical observation.

Today we know that measles and smallpox are transmitted from person to person through the air, in tiny droplets breathed out by infected victims. Although viruses can indeed survive for some time on inert objects before they infect another victim, this is not nearly as effective as direct droplet transmission. Once European diseases reached the American continent, they needed little help in transmitting themselves and often moved ahead of the European invaders. For example, in 1616–1617, a major epidemic devastated the Massachusetts Bay area. The Pilgrims did not land until three years later, by which time the indigenous population had already been thinned out.

Slavery and African diseases

One result of the massive die-off of the indigenous Americans was a shortage of cheap labor. The colonial powers responded by importing slaves from the Old World, who were consequently relatively resistant to Old World diseases.

In practice, this meant enslaved Africans who brought a variety of tropical diseases with them, notably malaria and yellow fever. These two diseases caused havoc in the tropical parts of the New World and are still major problems today in parts of South America. Of course, these diseases were also current among the Europeans and would have reached America whether or not slavery was practiced.

It is sometimes suggested that many human diseases originated in Africa. Doubtless some original human diseases (such as malaria, typhoid, and herpes) came from Africa because Africa is the home of the human species. However, until relatively recently, the population of sub-Saharan Africa was rather sparse. It is unlikely that most Old World epidemic diseases originated in sub-Saharan Africa. The dense populations of Mesopotamia, the Indus Valley, the Yellow River Valley in China, and the Nile Valley are more likely sites of origin.

Exposure of islands to mainland diseases

The islands of the Pacific Ocean were isolated from the rest of Eurasia, and not surprisingly, the various European voyages of discovery had much the same effect on the natives of the Pacific islands as on the Native Americans. For example, in 1875, measles reached the island of Fiji. Between 25% and 30% of the population died. Because individual islands are scattered over vast expanses of ocean, the smaller and lonelier ones have often remained uninfected until recently. They were usually exposed to one new disease at a time, giving their populations time to recover between onslaughts.

The islands of Britain and Japan were both relatively isolated from the neighboring continental mainland in early medieval times. Only after more efficient ships were developed and trading increased significantly did they reach infectious equilibrium with their respective mainlands. Consequently, England and Japan both experienced major die-offs due to epidemics brought from the disease-ridden mainland. Their population densities lagged well behind that of the mainland until relatively late in the Middle Ages. The Japanese population crossed the "epidemic threshold" and doubled between 1100 and 1300, whereas in England, this was delayed until after the Black Death of the mid-1300s.

Cholera and good intentions

Moralizers often blame the spread of epidemic disease on imperialism and exploitation. However, neither infectious agents nor their insect vectors subscribe to human ethics, and the spread of infections is neither hastened nor hindered by the moral (or immoral) intentions of the humans who carry them. Consider the spread of cholera. During the first worldwide cholera pandemic in the early nineteenth century, British ships carried cholera, which originated in India, to Muscat in Arabia. The British landed an expeditionary force there in 1821 to suppress slavery. From Muscat, cholera spread to the East Coast of Africa with the retreating slave traders. It then spread to the Persian Gulf and much of the Middle East.

Most contact between groups of people is the result of trading rather than the two extremes of invasion or charity. Regardless of human motivation, the spread of viruses such as smallpox, measles, and influenza merely requires close human contact. If Cortez had sailed to America to trade with the Aztecs, he would still have carried smallpox, measles, and influenza. If an international charity had sent Cortez to provide aid, the viruses would have gone, too. The overall demographic results would have been much the same in all cases.

The issue of biological isolation

Various authors have suggested that typhus and tuberculosis were present in both the New World and the Old World before contact. If so, the ancestors of these infectious agents must have crossed the Bering Strait with migrating humans around 10,000 years ago and then been isolated along with their human hosts. This generates a major biological paradox because microorganisms evolve rapidly. After 10,000 years of isolation, the New World and Old World descendents of any microorganism should have differed significantly in DNA sequence—and probably in clinical symptoms, too. Indeed, the Old World and New World species of trypanosome that cause sleeping sickness and Chagas disease, respectively, are quite distinct. Trypanosomes are protozoan parasites that evolve much more slowly than the bacteria responsible for typhus and tuberculosis. From an evolutionary perspective, it is extremely unlikely that descendents of such bacteria could remain indistinguishable if isolated for 10,000 years on different continents.

Such biological considerations imply that typhus and tuberculosis might have been present before contact in either the New World or the Old World, but not both. If the continents were truly isolated, we are therefore forced to choose one homeland for each of these infections. But just how isolated were the Eurasian and American continents from the perspective of infectious microorganisms?

Flocks of birds routinely migrate across the Atlantic Ocean. The microbiological consequences of this were recently illustrated by the appearance of West Nile Virus in New York in 1999. West Nile Virus infects a wide range of birds, among which it is spread by blood-sucking insects. Within the United States, it is spread by mosquitoes. The virus is widely distributed in Africa and the Middle East, but it had not previously been seen in the Americas. Birds carried the virus across the Atlantic, and DNA analysis suggests an origin among domestic geese in Israel. The issue is complicated by the modern possibility that birds sometimes rest briefly on ships, and insects can be carried accidentally by aircraft.

Spotted fevers and rickettsias

According to the records of the Aztecs and Incas, spotted fevers of some sort were present in precontact America. Degenerate bacteria known as rickettsias cause this group of closely related diseases. Rickettsias are much smaller than typical bacteria. They have degenerated and become so dependent on the host cells they infect that they cannot be grown in culture as normal bacteria can. (Note that the bone disease rickets has nothing to do with rickettsias. Rickets results from vitamin D deficiency, not infection.)

Rickettsias are spread by lice, fleas, or ticks. They are primarily infections of animals, transmitted to humans when their carriers jump from animals to nearby humans and bite them. Two major groups of rickettsial spotted fevers exist, each containing several closely related infections that are difficult to tell apart clinically because all cause spots and fever. The most famous is typhus, best known for devastating Napoleon's armies. The typhus fevers are generally regarded as Old World diseases, whereas the other spotted fevers, such as Rocky Mountain spotted fever, are of New World origin. However, things are not as clear-cut as they might seem.

Murine typhus primarily infects mice and other rodents. It is spread by fleas and lice, which can pass on murine typhus to humans by jumping or crawling from rodent to man. Epidemic typhus is a derivative that has adapted to humans. It is carried by human lice. Sophisticated diseases (such as malaria and yellow fever) do not usually kill their insect vectors. Similarly, murine typhus does not kill the fleas or lice that transfer it. In contrast, epidemic typhus does kill human lice, suggesting that epidemic typhus is a relatively novel variant, not yet fully adapted to its carriers.

The origins of typhus are uncertain

Although earlier reports indicate epidemics that might or might not have been typhus, the first unambiguous outbreaks were in the early 1500s, when the French and Spanish were fighting for control of Italy. From the beginning, typhus was associated with military campaigns and seems to have taken a clear dislike to the French, in particular. In 1528, the French were forced to withdraw from Naples after losing 30,000 men to typhus, which they presumably caught from the Spanish. From a focus in Spain, typhus spread around Europe. According to Spanish author Joaquin Villalba, the typhus epidemic of 1557–1570 depopulated much of the Iberian Peninsula. The timing and location supports the view that typhus might have been a New World import.

Supposed outbreaks of "typhus" in the 1400s in Germany or France are sometimes mentioned. In England, an outbreak of jail fever in 1444 killed 5 jailers and 64 prisoners. In recent centuries, "jail fever" has come to mean some version of typhus. However, back in the good old days, before different diseases were distinguished, jail fever was just that—what people caught in jail.

The Chinese medical classic the *Zhouhou Beijifang* lists a variety of diseases present in the Far East in the fourth century A.D. Among these is "Japanese river fever" or "sand-lice disease," often identified by modern commentators as tsutsumagoshi fever, or scrub typhus. The symptoms given are fever, bodily aching, and a rash. Several diseases fit these symptoms, but, if anything, the severe aching suggests dengue fever, which is widespread in the Far East and spread by mosquitoes.

In 1576, typhus, referred to as *cocoliztli*, swept Mexico, killing an estimated two million people. Supposedly, the Spanish rarely died. This does not square well with Villalba's claim that the 1557–1570 typhus epidemic depopulated the Iberian Peninsula. Nor does the higher death rate among the Aztec population support an American origin for this disease. However, we should remember that typhus is typically a disease of colder climates that relies on dirty clothing to harbor the lice that carry it. If typhus did come from America, it probably came from the colder, mountainous Inca lands instead of Central America. Thus, for the people of Mexico, typhus might well have been a new disease, brought back from Peru by the Spanish.

Although we cannot be sure, it seems likely that typhus and the related spotted fevers originated in the New World. Upon contact, typhus traveled back to Europe with the Spanish. In Europe, it increased in virulence over the next couple centuries as warfare provided it with a convenient means of transmission. Although more diseases have traveled from the Old World to the New, typhus and syphilis (see Chapter 7, "Venereal Disease and Sexual Behavior") appear to have been acquired in exchange. Admittedly, this is hardly fair trade, but virulent infections did not all travel in one direction, as is often suggested.

What about the Vikings?

In reality, Columbus was not first to discover America. The Vikings beat him to it. In 981, Eiríkr Thorvaldsson, better known as Eric the Red, explored Greenland. Eric's father was banished from Norway and went to live in Iceland. Eric himself was banished from Iceland and so set sail westward to explore. He found Greenland. After returning to Iceland, he led an expedition of 25 ships to Greenland because it had better grazing land than Iceland. Eric the Red apparently named the island Greenland to attract settlers. From Greenland, Leif the Lucky, son of Eric the Red, sailed to Newfoundland on the American mainland. The Vikings referred to the mainland as Vinland, and they settled there for about three years before they were driven away by the American Indians, whom they referred to as "Skrælings." Although Vinland was visited from Greenland several more times, no further settlement was attempted.

Notice that the Skrælings were not annihilated by diseases that the Vikings brought from Europe. This contrasts sharply with what happened when the Aztecs met the Spanish. Does this imply that smallpox, measles, and influenza were not current in Europe during this period of history? An alternative explanation is that the population of Iceland, the staging post for these voyages, was too small to maintain such diseases in circulation. Any epidemics brought from the European mainland to Iceland burned themselves out. Hence, the Vikings were relatively germ-free when they set sail and consequently failed to wipe out the Skrælings and colonize North America.

Although it seems clear that the Vikings did not bring any spectacular epidemic virus diseases to America, they may have brought other, less dramatic infections. Did they carry typhus? Typhus is spread by lice that inhabit human hair and clothes. Dirty clothes that are not changed or washed for the duration of an ocean voyage are ideal for lice to live and lay their eggs in. Because fur coats provide better homes for lice than loincloths, typhus is typically a disease of the colder regions. If typhus was extant in Europe at the time, it is hard to imagine the Vikings failing to take it with them. Here, then, we have an alternative scenario for typhus as a European disease that the Vikings carried to America before the Spanish invasion.

The Viking settlement in Greenland survived until the second half of the fourteenth century. Planetary cooling during this period made Greenland progressively colder and less fertile. But the final blow was plague. The Black Death of the mid–fourteenth century killed around 50% of the people of Scandinavia and Iceland. From there, it hit Greenland, probably in the winter of 1350. The details are unknown, but when the Norwegians visited Greenland in the early 15th century, they found only deserted villages. Presumably, the Greenlanders were wiped out before they could pass the Black Death on to the Skrælings.

10

Urbanization and democracy

Cities as population sinks

The growth of urban civilization conceals a major paradox. Clearly, the formation of the city-states of the Middle East and ancient Greece is good evidence for increasing population. The urbanization of Western Europe was also accompanied by an overall increase in population. What is rarely appreciated is that, until relatively late in the industrial era, the death rate in the cities was higher than the birth rate. Consequently, the populations of medieval cities were replenished by migration from the countryside. During most major periods of city growth, the population increases were actually produced in rural areas; the cities themselves had a net negative impact on the population.

The death toll in the towns was largely the result of infectious disease. Infant mortality was often greater than 50%, and many of those who survived infancy failed to reach adulthood. In medieval Europe, overall life expectancy ranged

from the mid-twenties to the upper-thirties. As industrialization proceeded, life expectancy rose and infant mortality fell.
However, even in England, it wasn't until the eighteenth century that most towns no longer needed rural immigrants to
maintain their numbers. Rural areas in medieval times were
healthy only in the sense that they generated a net increase in
population. Infant mortality and life expectancy were not as
abysmal as in the towns, but they still were horrific by modern standards. Many—perhaps most—of the peasants who
survived infancy died of fungal infections of the lungs, caught
from spores infecting their crops. In contrast, the towns were
home to the bacterial and viral diseases with which our modern city-oriented culture is more familiar.

Viral diseases in the city

The growth of dense populations and their crowding into
towns and cities allowed the emergence of highly contagious
viral diseases that circulate only among humans, such as smallpox, measles, mumps, and rubella (or "German measles"). By
industrial times, most of these had become childhood
diseases, causing very few deaths. Even smallpox, though still
a significant cause of death, had become much less dangerous.
In 735–737, smallpox killed around 75% of those infected and
annihilated half the population of Japan. By the late nineteenth century, the death rate from smallpox itself (*Variola
major*) was around 20% of those infected, and a variant known
as alastrim, or *Variola minor*, with a death rate of only 1% to
2%, was spreading. Infection with *V. minor* creates immunity
to the more dangerous type of smallpox, and it seems likely
that if nature had been left to take its course, in a century or
two, *V. minor* would have displaced *V. major*, relegating smallpox to a childhood disease comparable to measles. What

actually happened was that human intervention led to the eradication of smallpox by vaccination. The last cases of natural smallpox occurred in Bangladesh, in 1975 for V. *major* and 1977 for V. *minor.*

Bacterial diseases in the city

Bacteria do not evolve as fast as viruses, so most new bacterial diseases of humans are still shared with other animals. We can divide the bacterial diseases of the growing urban populations into three main groups: those spread directly from person to person, those spread by dirty water, and those spread by insects. Bacterial diseases spread from person to person, usually by droplets coughed or sneezed into the air, were common in towns and cities until recent times. Examples include scarlet fever, diphtheria, whooping cough, and tuberculosis. These diseases spread in a manner similar to that of measles and smallpox, although less efficiently. Vaccination has largely eradicated most of these diseases from advanced nations.

We saw in Chapter 4, "Waters, Sewers, and Empires," how the collapse of the Indus Valley civilization was probably linked to cholera or a similar waterborne infection. Typhoid, dysentery, and a host of lesser diarrhea-causing diseases took a steady toll throughout the period following the Roman Empire, especially in crowded communities whose water supplies were at risk of contamination. As technology advanced, these diseases have gradually faded away in industrialized nations. Today various diarrhea-causing strains of *Salmonella* and *E. coli* have re-emerged as public health hazards, less in the water supply than in batches of contaminated food, especially processed meat.

The Black Death

The outstanding example of a bacterial disease spread by insects is the Black Death, or bubonic plague. In the middle of the fourteenth century, it wiped out around half the population of Europe. Very likely it did as much damage in the more densely populated parts of Asia, the Middle East, and North Africa, although detailed records are available only for Europe. Lesser outbreaks of bubonic plague reverberated around Europe until the 17th century.

Bubonic plague is caused by the bacterium *Yersinia pestis,* which infects many animals, especially rodents. From these, it can be transmitted to humans (and their cats and dogs) by fleas. Why would fleas leave the shelter of the densely packed hairs on furry animals such as rats to venture onto the exposed surface of relatively hairless creatures such as humans? Not through choice. Fleas come in distinct varieties and tend to stay with the animal they are adapted to. But if the animal dies, the flea can no longer obtain its required diet of fresh blood and must find a new host. So when rats or mice die of plague, their fleas leave and look for new animals to infest. Rat fleas cannot actually survive for long on humans—our blood doesn't supply the correct balance of nutrients. But one bite is enough to transmit plague. That the flea will eventually die due to improper nutrition is small consolation.

In nature, the plague bacterium infects wild rodents such as the marmots and susliks of central Asia, or their relatives, the ground squirrels and chipmunks, of North America. It causes only mild symptoms, often no worse than a bad cold would be to humans. When displaced from their normal environment, the wild rodents can transfer bubonic plague to the rats and mice who live in close contact with humans. Unlike their wild relatives, domestic rats and mice fall seriously ill and are killed by plague. Their resident fleas then look for new animals to live on. This is what sets in motion epidemics of plague. The great Black Death epidemic of the Middle Ages was probably the result of climatic fluctuations in the northern Asian steppes. A few good years—for rodents—followed by a couple bad years resulted in a large rodent population with no food. So, several rodent species extended their ranges southward. This brought them into contact with other, more southern, rodents that, in turn, made contact with the human societies of Asia. The fleas and *Yersinia pestis* bacteria were passed along, too.

Climatic changes: the "Little Ice Age"

Before plague struck Europe in 1347, things had been going downhill for the better part of a century. Although fourteenth-century Europe had 10% or less of today's population, it was overpopulated in the sense that the cultivated area and level of technology in use produced barely enough food. From 750 to 1200, things went reasonably well: Crop yields increased and the population grew. Between 1200 and 1350, Europe got colder and wetter. Pastures high in the Swiss Alps were covered under glaciers and did not re-emerge until centuries later. The Thames River in England froze over a dozen times in the 1400s. During this "Little Ice Age," the weather in Europe was probably the worst since the Great Ice Age of prehistoric fame.

From about 1250 onward, Europe began to spiral downward into poverty. Crop yields decreased. An expanding population forced greater reliance on a single staple crop—wheat in Europe instead of rice, as in Asia. Malnutrition spread as diets became less varied. Crop rotation was often suppressed, and fallow land was planted in cereals in an effort to feed more people with a minimal diet. This exhausted the land, and yields dropped further. From about 1290 until the Black Death arrived in 1347, crop failures of increasing severity caused frequent local famines over much of Western Europe. Starvation, accompanied by intestinal infections such as typhoid and dysentery, reduced Europe's population by 10% to 20% during this period. And then from the steppes of Asia came the solution to the European population problem: the Black Death.

The Black Death frees labor in Europe

Human populations are remarkably resilient, and if the epidemic of 1347 had been the only outbreak of plague, human numbers would have soon recovered, even from a 50% death toll. The real horror of the Black Death was that epidemics recurred constantly for the next 300 years. The first hundred years was by far the worst, with every generation hit by plague. Depopulation continued for more than a century. Gradually, the epidemics faded and population recovery began a century or so before the Black Death disappeared. In the short term, Europe was totally devastated. However, we must distinguish between short-term catastrophe and beneficial long-term effects. Much has been written on the effects of the Black Death, so we need not go over the evidence in obsessive detail. Several major long-term benefits are generally recognized.

The population collapse of the fourteenth century halted Europe's spiraling descent into famine and poverty. Simply put, fewer people meant more farmland and, therefore, more food per person. It is true that, early on, the disruption caused by the Black Death resulted in local famines. But these were short-term effects of dislocation. After society adjusted, the benefits of a decreased population became evident. The resulting scarcity of labor meant that laborers, especially skilled craftsmen, became more valuable. Although governments attempted to control wages and prices on behalf of landowners, they failed. Peasants who were ill-treated, however legally, simply moved. Labor was in such demand that employers and landlords who benefited from new arrivals could be relied on to turn a blind eye to regulations prohibiting the movement of labor. Mobility of labor and higher wages led not only to a higher standard of living for most ordinary folk, but also to a more market-oriented economy overall.

The same labor shortage also led to greater interest in ways to increase output despite less human labor. More mechanization, better technology, and the emergence of the experimental approach, which underlies modern Western science, all trace their origins to this era. This is not to say that no technical advances occurred before the Black Death. Some did, but surplus manpower made labor-saving devices of little importance. Depopulation merely shifted the balance in favor of technology.

Death rates and freedom in Europe

The Tatars brought plague from the Asian steppes to the Crimea. From there, plague was carried around the Mediterranean to Europe, the Middle East, and North Africa. Plague landed in Europe in 1347. Death rates throughout Europe ranged from a mere 25% in luckier areas to 60% or more in the worst-hit regions.

One vivid quotation from Florence, one of the hardest-hit of the Italian cities, shows the horror:

"At every church, or at most of them, they dug deep trenches, down to the waterline, wide and deep, depending on how large the parish was. And those who were responsible for the dead carried them on their backs in the night in which they died and threw them into the ditch, or else they paid a high price to those who would do it for them. The next morning, if there were many [bodies] in the trench, they covered them over with dirt. And then more bodies were put on top of them, with a little more dirt over those; they put layer on layer just like one puts layers of cheese in a lasagna."—Stefani, Marchione di Coppo (written late 1370s/early1380s)

According to Boccaccio, more than 100,000 died in Florence in 1347–1348. Modern scholars usually accuse Boccaccio of exaggerating, because Florence is estimated to have had a population of about 80,000 at this time. Modern estimates of the death rate in Florence range from 45% to 75%. Although ancient writers often exaggerated, modern commentators suffer from the opposing bias. In our own vaccinated and disinfected era, it is difficult to imagine that so many people could possibly succumb so fast to infectious disease.

A key unanswered question is the accuracy of historical estimates of population. In the most recent American census, a substantial number of inner-city slum-dwellers went unrecorded. By what percentage the U.S. population was underestimated is still debated. Perhaps more pertinent is the situation in the overpopulated cities of the Third World. Does Manila or Mexico City have an accurate roll of the slum-dwellers who live in the shantytowns around their perimeters? If Ebolavirus swept Calcutta next year, the body count might well be greater than the official population.

Although mortality was usually higher in cities, the plague also tended to mutate from the bubonic form to the more deadly pneumonic version in colder, damper regions. Thus, Scandinavia was very hard hit, despite its relatively low population density. Eastern Europe was last to be attacked, in 1350–1351, and, for reasons unknown, had the lowest death rates. It is interesting to note that the regions hit the hardest by plague were those where freedom and Western technology eventually developed the fastest. Remember that, up to this point in history, the peak of European culture had been centered around the declining Byzantine Empire in the East. In contrast, Western Europe was relatively backward. After the Black Death, this reversed.

Eastern Europe, then, was luckier in the short term but not so lucky in the long term. Its death toll from the Black Death was half that in the west. Consequently, the disruption and manpower shortages were less overwhelming. Aristocratic landlords and the Church were able to maintain control of most of Eastern Europe, with the result that it remained impoverished and relatively backward until the twentieth century. The long-term benefits of the fourteenth-century population collapse are perhaps most evident when we compare the prosperous, technologically advanced societies of Western Europe with the backward, poverty-stricken countries of Eastern Europe. Technology, prosperity, and religious freedom are all inextricably intertwined in the formation of modern industrial democracy.

The Black Death and religion

The effect on religion was paradoxical. It was inconceivable to the medieval mind that the death of so many people in just a few years could be anything but a sign from heaven. If anything, belief in God was strengthened, yet clearly something was wrong. The high mortality among ordinary parish priests indicates that most remained loyal to their flocks during the Black Death. However, the Church as a whole lost authority and respect during this period. Before the Black Death, some 25% of all bequests in wills had been to the Church. After the plague, instead of donating money to the Church, pious merchants and nobles founded private charities, thus removing large sums from the control of the Church. In addition, much of the Church's wealth was in the form of land. Higher wages and lower food prices made land owning far less profitable than it had been in the days before the population collapse. Thus, loss of respect was followed by loss of income, and the Church's influence dwindled.

The ever-increasing tendency to cut out the religious establishment and deal with God on a direct personal basis led to demands for Bibles in local languages and to revolts against the papal religious monopoly. To compound matters, the papacy was in disarray during this period. Indeed, from 1378 to 1415, there were two competing Popes, one in Italy and the other in France. This did little to improve the image of the Church. In England, John Wycliffe (1324–1384) and the Lollards rejected the authority of the Church and instead emphasized the Bible. Wycliffe, a professor at Oxford University, translated the Bible into English and also preached against the Catholic establishment, especially the Dominican and Franciscan monastic orders. According to Wycliffe, "Friars draw children from Christ's religion into their private Order by hypocrisy, lies, and stealing." Canon Knyghton of Leicester replied by claiming that Wycliffe was casting the Gospel pearl under the feet of swine by making the Scriptures available in English "to the laity and to women who could read." The eventual outcome of the ever more widespread religious dissent was the Reformation. Protestantism spread throughout northwestern Europe, and this led eventually to religious freedom.

The White Plague: tuberculosis

The last epidemic of bubonic plague in northwestern Europe was the Great Plague of London in 1665. The last outbreak in the western Mediterranean region was in Marseilles and the surrounding area of France in 1720–1721. These late outbreaks were not only restricted in area, but also had lower mortality rates than outbreaks earlier in the pandemic. The Black Death was over, and the burden of controlling Europe's growing population fell briefly to smallpox and then to a new bacterial disease, tuberculosis. Victims of bubonic plague developed purple-black patches on the skin. In contrast, the victims of tuberculosis were deathly pale and wasted away slowly. So tuberculosis was sometimes called the White Plague, in contrast with the Black Death. More often it was called consumption because of the slow wasting away.

Malnutrition, especially a shortage of protein, lowers resistance to tuberculosis. This was a major factor in the rapid spread of tuberculosis among working-class children during the eighteenth and nineteenth centuries. The healthy body attacks tuberculosis bacteria that are invading the lungs in two ways. It generates toxic nitric oxide in the lung tissue and walls off any surviving bacteria behind layers of immune cells. Both responses are weakened by a low-protein diet. The effects of tuberculosis in Africa illustrate this effect. In the 1930s, the Kikuyu tribe of East Africa suffered major losses from tuberculosis, whereas the related and neighboring Maasai suffered much less. The Kikuyu were largely vegetarian farmers, but the Maasai cattle herders lived on meat, milk, and blood.

Tuberculosis was responsible for about 20% of all deaths in England in 1650. Its share of deaths declined during the next century, probably due to the prevalence of smallpox in this period. Tuberculosis was back as the leading killer by the early 1800s. From the mid-1800s, the death rate from tuberculosis declined steadily until the 1950s. At the start of the twentieth century, essentially all city dwellers in Europe and America tested positive for tuberculin, indicating that they had all been infected. Yet few had active tuberculosis, indicating that most of the population of Europe was by then resistant. In 1926, a tragic mix-up in Lubeck, Germany, resulted in 249 babies being injected with virulent tuberculosis instead of the vaccine strain. Only 76 died, indicating 70% resistance, even among infants whose immune systems were still not fully operational. Starting in the 1950s, the discovery of antibiotics almost eradicated tuberculosis from the industrial nations.

The rise of modern hygiene

As the nineteenth century progressed, technology and industrialization ushered in modern hygiene. Clean water, flush toilets, sewers, and soap united with better nutrition and improved housing to vastly reduce the incidence of infectious disease. Mass production of cheap cotton underwear that was easy to clean helped things along. The knowledge that germs caused disease led to changes in acceptable human behavior. Samuel Pepys's diary, written in the 1660s, tells us this:

> I was sitting behind in a dark place, a lady spit backward on me by mistake, not seeing me, but after seeing her to be a very pretty lady, I was not troubled at it at all.

Today spitting is no longer polite—even by attractive women, even on the floor.

During the twentieth century, vaccinations and antibiotics finished the job civil engineering had begun. Infant mortality shrank to a vanishing point. Today the inhabitants of industrial nations expect to die of heart failure or cancer only after surviving the threescore years and ten of traditional wishful thinking. Smoking has joined spitting on the list of proscribed behaviors. Yet overeating, a far bigger threat to health, has received remarkably little bad press in comparison.

Although it is certainly better, on average, to be scrubbed, shampooed, disinfected, and vaccinated, there is a downside to the overexuberant use of hygiene. Perhaps the first example to be noticed was the curious case of polio. Starting in Sweden in the 1880s, poliomyelitis increased in frequency in countries with increased levels of hygiene. It remained virtually absent in countries with preindustrial standards of sanitation. Most cases of infection with poliovirus show no noticeable symptoms. A minority suffer mild fever and diarrhea. In a tiny proportion, the virus penetrates beyond the intestines and attacks the nervous system, resulting in paralysis of the lower limbs. These unlucky few attracted attention to the disease.

Polio is an intestinal disease. It is a member of a large family of closely related viruses that are all passed on by contamination of water with human waste and that infect the lining of the intestines. These viruses cause mild intestinal upsets, usually so mild there are no visible symptoms. In unsanitary societies, all children are infected with members of this virus group early in infancy and so become immune. Because these viruses are closely related, immunity to one gives partial or total immunity to others. Consequently, children raised in unhygienic conditions are almost always at least partially immune to polio before infection. As hygiene improves, infection by viruses in this family decreases sharply. If children now contracted polio, they would be unlikely to have been preimmunized by one of the less virulent relatives. The result would be a more aggressive infection, occasionally resulting in the tragedy of paralysis. Fortunately, artificial immunization has now largely eliminated polio from advanced nations, and major progress has been made toward worldwide eradication.

Despite such relatively minor setbacks, by the mid-twentieth century, the combined effects of antibiotics and immunization, superimposed on earlier advances in civil engineering and hygiene, had all but eliminated infectious disease from the advanced nations. Although there have been some reversals since, this is undeniably one of the greater triumphs of scientific man.

The collapse of the European empires

Today's industrial nations got modern hygiene and health care after industrialization. Consequently, infectious disease kept their populations in check until the industrial revolution was underway. In contrast, most Third World nations received the basics of modern hygiene and health care from the advanced nations before industrialization. Hence, their population growth was checked much less by infectious disease.

The resulting massive population growth undermined the prosperity of the European colonial empires. Thus, the British Empire had the opposite problem from the Roman Empire. The Romans ran out of manpower, whereas the British were overwhelmed by surplus population. Thus, today's major divide between the advanced nations and the

Third World is, to some extent, a legacy from the effects of overcrowding and infection.

Resistant people?

From the 1300s until the 1900s, the European population was brutally culled by bubonic plague, smallpox, and tuberculosis, to name just the three main culprits. As noted earlier for TB, many of the survivors apparently had genetic alterations that made them resistant. But exactly what has been selected?

We know of some genetic alterations that confer resistance to a specific disease or group of related infections. As already discussed in Chapter 4, alterations in the cystic fibrosis gene protect against infections that cause dehydration due to diarrhea. A variety of mutations are known to protect against malaria (see Chapter 2, "Where Did Our Diseases Come From?"), although these would not affect urban dwellers in the temperate zone. The $CCR5\Delta32$ mutation is found in about 10% of Europeans and confers resistance to AIDS. It has been suggested that this mutation was selected by plague or smallpox. However, recent data shows that it was present with the same frequency in Bronze Age Europeans some 3,000 years ago, well before plague or smallpox were prevalent. Thus, the selection force behind the $CCR5\Delta32$ mutation is still a mystery.

Two major possibilities that might protect against multiple infections are altered behavior and changes to the immune system. Being smart enough—or perhaps just cowardly enough—to run away when plague threatens increases the likelihood of survival considerably. Even during the Black Death, some isolated villages escaped almost unscathed. Intelligence correlates with prosperity, and in medieval times those who were richer tended to sleep with fewer companions—both human and rodent. Records confirm that the urban poor who often slept a dozen to a room on a bed of straw suffered more casualties than the prosperous.

Modern genetic data suggests that at least a thousand or so of our genes are expressed at especially high levels in the brain. Deciphering which genes affect which aspect of brain function or behavior and whether they show signs of recent evolutionary selection is a daunting task. Currently, we can do little more than speculate. Personally, however, I find it difficult not to believe that massive epidemics have selected at least to some extent for increased intelligence, as well as behavior that reduces the likelihood of infection. Elevated caution (or cowardice) and increased preference for a solitary existence should also have been favored.

How clean is too clean?

Another possibility for broad resistance to several infections is a more aggressive immune system. Our immune system must be carefully balanced. If the immune system is too cautious in reacting, infections may win; if it is too trigger happy, we can damage our own tissues. Clearly, the optimal setting depends on the likelihood of encountering dangerous infectious diseases. Thus, urban plagues might have favored an overenthusiastic immune system. Although this was beneficial at the time, its legacy could be an increased level of autoimmune problems. These range from allergies and asthma to arthritis and multiple sclerosis. Autoimmune problems are most prevalent in industrial nations where overcrowding was worst; they are much rarer in Third World populations.

Another viewpoint on the higher frequency of autoimmune problems in advanced nations is the level of hygiene. Eating dirt is not generally recommended in manuals on the care of babies and small children, but mounting evidence suggests it might not be such a bad idea. When children are fed sterilized food, given too many antibiotics, and vaccinated instead of gaining immunity from natural infections, the immune system develops in a lopsided manner. This correlates with the increasing frequency of immune system disorders in industrialized nations. Children who develop immunity by natural exposure rarely suffer from these afflictions. Moreover, the culprits favored by tradition—pollution, dust mites, toxic chemicals, and colds—have now largely been exonerated. Although it would be silly to expose infants to sources of dangerous infection, some sort of exposure to "clean" dirt might not be a bad thing.

Where are we now?

For better or worse, we are living under artificial conditions far different from those of our hunter-gatherer ancestors. We have also been genetically modified in ways that we are just beginning to glimpse. So, what of the future? We consider our future conflict with infection in the final chapter.

11

Emerging diseases and the future

Pandemics and demographic collapse

Today our planet holds approximately 6 billion humans and 5×10^{30} (5 million trillion trillion) bacteria. We are outnumbered by nearly 10^{21} (1 sextillion) to 1. At the moment, we are catching up. But will this trend last? The human population does not climb smoothly. Periods of growth are followed by population crashes. When will the next population implosion happen? How?

The earliest major population collapse for which we have reliable records occurred in the Roman Empire as a combined result of the unidentified pestilences of 165 A.D. and 251 A.D. The plague of Justinian, which began in 542, with secondary epidemics until 750, was another period of major population decline in Europe and the Middle East. Both population and prosperity increased from the mid-700s to the mid-1300s. In England, the population expanded roughly threefold from 1000 A.D. to 1348, to reach roughly four million. It then collapsed to less than half of this due to the Black Death and regained its 1348 level only in the early 1600s, some 250 years later. Europe, North Africa, the Middle East, and China suffered similar catastrophic die-offs during the same period. In the New World, these multiple die-offs combined into one spectacular population crash when the diseases of the Old World were imported into the Americas, beginning in 1492.

Since around 1600, despite the ravages of tuberculosis and other infections, the population of most parts of the world has steadily increased. Today the threat from infectious disease is growing. The industrial nations are shielded by wealth and technology from the infections that assault the poor nations. Yet despite the poverty, crowding, malnutrition, and lack of hygiene, the populations of the Third World nations continue to rise. Whether we are likely to succumb to some new plague in the near future and suffer another major population collapse is hotly debated.

The various types of emerging diseases

The widespread publicity given to AIDS, mad cow disease, and Ebolavirus has set a trend. Virtually every infection now clamors to be accredited as an emerging disease. Diphtheria is increasing in the former Soviet Union, and syphilis is increasing among American homosexuals. Are these emerging diseases? Not really. Localized lapses in social order or hygiene provide familiar diseases with the opportunity to briefly expand. However, the idea of emerging disease implies something truly novel that threatens a world grown smug in the belief that infectious disease has been conquered. Changes in one or more of four major areas can qualify a disease as novel.

Genuine emergence of a novel infection is rare. Many so-called emerging diseases have survived unnoticed for many years and only recently come to our attention as a result of changing conditions. Others are truly novel and have emerged by a combination of genetic changes and movements between host species, the most clear-cut case being AIDS.

Changes in knowledge

Many "novel" diseases have clearly been around for a while but were only identified recently. Previously unnoticed diseases include Legionnaire's disease and Lyme disease. A hundred years ago, deaths from Legionnaire's disease would have been attributed to tuberculosis. The decline in tuberculosis means that rare diseases affecting the lungs are more likely to be noticed. Similarly, the decline in yellow fever, due to vaccination, has revealed many less common tropical fevers that were once lumped together with yellow fever. Conditions such as diarrhea or hepatitis that have been known for a long time are caused by multiple infectious agents. Many of these have only recently been individually identified, including hepatitis viruses C to G.

Other diseases went unnoticed because they occurred only in out-of-the-way places or among low-status people. Frontiersmen and American Indians have undoubtedly suffered from sporadic cases of Lyme disease for centuries, but the disease was investigated only when it began to affect leisured landowners.

Changes in the agent of disease

Totally novel diseases such as AIDS and mad cow disease appear from time to time. These infectious agents are of recent origin, and there has never been a recorded outbreak before our own time. Mad cow disease, though bizarrely unique, causes few deaths. Its main effects have been economic.

As agents of a catastrophic die off, these two diseases share the same drawback: ineffective transmission. If AIDS were spread by insects or mad cow disease wafted through the air like measles, we would be in real trouble. Perhaps most diseases with efficient distribution mechanisms are already in circulation. Diseases still lurking in the shadows are probably obscure for a good reason—they can't effectively transmit themselves to humans. Although novelty dominates the headlines, I believe that greater real danger comes from known diseases that transmit themselves efficiently. If measles or flu changed to become highly virulent, we might face a real possibility of a major die-off.

Old diseases can evolve to protect themselves against human counterattack. They might gain resistance to antibiotics or change their surface components to outwit the immune system. Influenza virus changes its surface properties for each new epidemic, and AIDS virus mutates so rapidly that multiple variants appear within a single patient.

Old diseases can evolve new means of attack. They might acquire novel virulence factors (such as with *E. coli* O157) or reshuffle their genes, creating virulent variants from time to time, as occurs with flu. Other diseases venture into new territory by changing the tissue invaded. A good example is the evolution of the spirochete that causes the skin disease yaws into syphilis, which specifically affects the genital regions.

Changes in the human population

Humans with defective immune systems provide easy opportunities for infection. The AIDS epidemic has created by far the largest number of immune-compromised victims. This, in turn, has allowed the spread of many novel opportunistic diseases. However, the use of drugs whose side effects harm the immune system and the increasing numbers of older people also contribute. Malnutrition also makes humans more susceptible to many infections. Denser human populations allow more virulent variants of a disease to spread, as discussed in Chapter 3, "Transmission, Overcrowding, and Virulence." This is especially true of the growing cities of the Third World, where poor hygiene exacerbates the effect.

Genetic alteration of humans might protect against one problem but increase vulnerability to others. A single copy of the cystic fibrosis mutation protects against diseases such as typhoid and cholera, but if both copies of the gene are defective, the lungs become more susceptible to infection by a variety of bacteria.

Multiple genetic changes have occurred within historical times that make most Europeans relatively resistant to smallpox, measles, tuberculosis, and many other diseases. Because we do not know the identity of most of these mutations, we remain in the dark about any possible side effects.

Changes in contact between victims and germs

A variety of factors can greatly change the probability of an infectious microbe finding suitable victims. Natural disasters such as earthquakes, floods, hurricanes, and storms provide temporary opportunities for disease to spread. Generally, when the disaster is past, associated infections fade away, too. Changes in climate have more permanent long-term effects. In particular, higher temperatures allow insects to spread into new regions, carrying with them diseases such as malaria or yellow fever.

Old diseases can make a comeback due to a breakdown in public health caused by wars, revolutions, and political upheavals. Disruption of vaccination programs after the collapse of the Soviet Union resulted in an upsurge of diphtheria.

Opening up new land for settlement or clearing forest for agriculture brings people into contact with previously unknown diseases, such as Ebolavirus. More serious in practice are irrigation projects that create new bodies of standing water. These permit the spread of mosquitoes carrying malaria, yellow fever, and dengue fever. Deforestation also helps the spread of disease-carrying insects. Even technological advances can backfire. More efficient food technology relies on processing larger volumes. This allows localized bacterial contamination to spread more widely, resulting in the massive meat recalls of recent years. Air-conditioning opened new opportunities for Legionnaire's disease.

The supposed re-emergence of tuberculosis

Tuberculosis is not an emerging disease. Indeed, it scarcely merits classification as re-emerging. We include it here because of the publicity it receives. Tuberculosis spread through the cities of Europe during the industrial revolution. By the mid-twentieth century, when the first effective antibiotics appeared, the tuberculosis epidemic was already largely burned out in Europe. A colossal death toll over the past few hundred years had killed off most people who were sensitive. Antibiotic therapy merely finished off the tail end of the epidemic in the industrial nations. Today only an estimated 10% of the white population is susceptible to tuberculosis.

Thus, tuberculosis is not re-emerging; the epidemic that started in seventeenth-century Europe did not end; it merely vanished from the advanced nations. Today it is moving inexorably across the world and is still expanding among populations who have not been previously exposed. Today tuberculosis accounts for around three million deaths annually, most in the Third World. Data from Chicago in the 1920s showed that tuberculosis was six times worse among blacks than whites. Remember that some 75% of American whites are descendents of Europeans who entered the United States between the American Civil War and World War I. The whites thus came from a pre-exposed population, whereas the blacks did not. After World War II, tuberculosis found fresh victims in the crowded slums of expanding Third World cities. Susceptibility is greatly increased by the protein-poor diets often found in poor countries where little meat is eaten.

Diseases are constantly emerging

Quite frequently, die-offs occur among wild animals. Most of these are not noticed. Most that are noticed, usually by local farmers or hunters, are not recorded, and most of those recorded are never explained. For example, in 1994, bald eagles in Arkansas came down with a mystery disease and many died. Some die-offs are probably the result of new diseases. Perhaps an existing infection mutates to a more lethal form, or perhaps a disease crosses over from another animal. This new disease is so virulent that it wipes out most of the population and then, with no more victims to infect, goes extinct itself. This undoubtedly happened many times to early human settlements in the days before the human population was dense enough to keep new diseases in circulation. For every disease that emerges successfully, many must make abortive attempts. Even if a disease ultimately emerges into new prominence or jumps into a new host, it might make many tries before it succeeds. Let's look at some recent candidates for fame and glory.

Lassa fever, Hantavirus, and Ebolavirus are three of the emerging viruses to hit the headlines recently. All three are harbored by small animals with large populations. Lassa and Hanta are carried by rodents, and Ebola most likely by bats. As with the Asian marmots that harbor the Black Death, the natural hosts for Lassa, Hanta, and Ebola suffer relatively mild disease. Man is very susceptible, and the death rates during the reported outbreaks are in the same range as for bubonic plague.

These three, together with Junin, Machupo, Marburg, O'nyong Nyong, and several other novel viruses, were mostly identified more than 25 years ago. However, it wasn't so much the viruses that emerged—it was the emergence of cell culture techniques for viruses in the 1950s that allowed the identification of exotic viruses in remote parts of the world. Before this, viruses had to be grown in fertilized chicken eggs, an extremely laborious and clumsy procedure that was not always successful. Few novel viruses of major significance have appeared in the last 25 years, although there have been new outbreaks of those listed earlier in new places (for example, Hantavirus in the United States in 1993). Most recently, Lujo virus, a relative of Lassa fever, emerged in Southern Africa in late 2008.

Despite the hype, these novel viruses have had little global impact. About 50 million people die each year on Earth. Of these, about 16 million succumb to infections, with AIDS, malaria, and TB accounting for roughly 3 million each. Deaths per year from the newly emerging viruses are numbered in the hundreds. For example, between its emergence in 1976 and the outbreak in 1996, there were approximately 1,000 official cases of Ebolavirus infection, with an overall death rate of 80%. Undoubtedly several thousand more victims died unreported in isolated villages; nevertheless, in global terms, these numbers are negligible.

How dangerous are novel viruses?

Should we be alarmed about these novel viruses? Worried, yes—panicked, no. Consider Lassa fever, discovered in 1969 in Lassa, Nigeria. Like Ebolavirus, Lassa fever virus has undoubtedly been around for much longer. Sporadic outbreaks in isolated areas must have occurred from time to time without drawing attention. Earlier outbreaks were probably misdiagnosed as severe cases of malaria or yellow fever. In its natural hosts, small rodents, Lassa fever often causes mild long-term infections. The virus emerges in the urine and can be breathed in by humans under dry, dusty conditions. In humans, Lassa fever is virulent and short-lived—as is the patient, in the majority of cases. Human survivors also excrete virus particles in their urine for up to a month after infection.

The story is similar for Ebolavirus, named after the Ebola River in Zaire. Bats can be infected with Ebolavirus, whereas many other animals, including rodents, cannot. As expected for the natural host, bats allow the virus to replicate but do not fall severely ill. Outbreaks of Ebolavirus in the Sudan were traced to a cotton factory whose rafters were home to thousands of bats. An outbreak in Uganda was traced to the bat-infested Kitum Cave. Despite this, no bats trapped near these human outbreaks actually carried Ebolavirus. So although bats are the chief suspects, the case remains unproven.

However, despite being highly lethal, Ebolavirus is not especially infective until the final stages, when the blood is full of virus and the patient bleeds from all the bodily orifices. Obviously, at this stage, the patient is immobile. Lassa fever and Hantavirus are much the same. Exposure to blood or tissue samples has infected health workers, but casual contact rarely transfers the virus. During the 1995 Ebolavirus outbreak in Zaire, of 28 relatives who stayed in the hospital to help nurse the sick (often sharing the same room or even the same beds), 17 got Ebola. Of 78 who just visited, none fell ill. Similarly, in 1990, an American returning to Chicago from Nigeria was hospitalized and died of Lassa fever. No one else was infected, although no special precautions were taken because the infection was identified only after he died.

Lassa, Ebola, and Hantavirus are not as evil as originally believed. Mild versions of all three viruses are surprisingly widespread. Investigations during the 1980s in the rain forests of Cameroon found that 15% of the pygmies had antibodies to Ebolavirus in their blood, implying that they had been infected. No massive death toll was noted. Similarly, screening of large numbers of Africans from Nigeria and nearby nations found many who had signs of having been infected with Lassa fever but remembered only mild illnesses.

Another factor contributing to the panic of the early Lassa and Ebola outbreaks was their artificial spread by hospitals. Thus, many victims of the 1976 Ebola outbreak in Zaire were infected while in the hospital. Viruses from patients infected with Ebola were transferred to others by reusing hypodermic needles that were improperly sterilized. Less than 10% of those who got Ebola injected directly into their bloodstream survived. Of those who caught Ebola from another person, between 40% and 50% survived. Similar scenarios are seen with Lassa fever. Thus, these diseases are less dangerous when spread by natural means.

Transmission of emerging viruses

It is sometimes suggested that highly virulent diseases cannot spread very far unless they are carried by vectors such as fleas or mosquitoes. Granted, milder variants of a disease tend to spread further. Nonetheless, if Ebolavirus or Lassa fever had started out capable of infecting humans efficiently, the diseases could have spread much further. We know this from what happened when smallpox and measles first reached the American continent.

So why have the outbreaks of Ebolavirus and Lassa fever burned out so rapidly? For a person-to-person disease, two factors affect transmission. First, how many other people does the infected victim contact? This depends on how long he survives and how far he can move. Clearly, this factor reduces the spread of more virulent diseases. Second, how well does the virus jump from one person to another upon contact? This is a property of the virus itself and varies greatly. As noted earlier, although both are highly virulent, neither Ebolavirus nor Lassa fever is especially infectious to humans.

Moreover, for every person exposed to deadly viruses such as Ebola or Lassa, a thousand are exposed to other unknown viruses from rats, bats, monkeys, zebra, elephants, and other animals. Most of these unknown viruses will never make the headlines because the body's immune system zaps them as soon as they set foot inside. In summary, humans could be highly susceptible to viruses they have never been exposed to, but most viruses that have not adapted to humans will be extremely susceptible to immune system destruction. Occasional strange viruses do escape the immune system and cause a great deal of damage. But because they are not adapted to humans, these chance invaders rarely have an effective way to move from person to person.

Efficient transmission and genuine threats

For a genuinely new human plague to emerge, the agent of disease must evolve (or already possess) some way to spread efficiently. Probably the best way is to spread through the air, directly from person to person, as with flu or measles. Despite killing only a tiny fraction of their victims, measles and flu both kill far more people than all the novel emerging viruses combined. This is because they infect vast numbers of victims.

Influenza virus changes both by rapid mutation and by exchanging genes between flu strains from people, pigs, and poultry. Most new flu strains have only minor alterations, but now and then, major changes occur that produce variants with increased virulence. In the twentieth century, this occurred in 1918 and, less impressively, in 1933, 1957, 1968, and 1977. The virulent flu of 1918 had an overall mortality of only 2% to 3%, but because it infected most of the human population, the total death toll was huge: 30 million to 100 million, according to different estimates.

In 1920, the world population was a little less than 1.5 billion, and in the year 2000, it was a trifle more than 6 billion. Thus, today's population is about four times as dense as during the Spanish flu of 1918. As we know, the denser the host population, the more this favors the spread of virulent infections. Influenza and assorted respiratory infections, ranging in mildness down to the common cold, are caused by hundreds of different viruses that constantly mutate. These viruses are well adapted to transmission among humans. Sooner or later, one of these, not necessarily flu itself, will likely throw up a virulent variant. As Pasteur might have put it, chance favors the prepared microbe.

The history and future of influenza

The first definite flu epidemics were recorded in Europe in 1510, 1557, and 1580. The mortality rates of up to 20% were vastly more lethal than any recent epidemic except the 1918 Spanish flu. Over the centuries, either flu has been getting milder or humans have been getting more resistant. The flu epidemics of the 1500s killed a substantial proportion of the population and must have weeded out many sensitive members of the population.

Flu is shared among people, pigs, and poultry. Most new variants of flu originate in China, where large numbers of ducks and pigs live close to humans. The Asian flu of 1957 and the Hong Kong flu of 1968 are typical examples. Direct chicken-to-human transfer of influenza is very rare. Generally, flu goes via ducks and pigs. However, in Hong Kong in 1997, a handful of people died from avian flu that was apparently transferred directly from chickens. Luckily, this flu virus did not transfer between humans. It's unlikely that future outbreaks of flu will result from direct fowl-to-human transfer of a virus that is both lethal and capable of spreading from person to person, but this is still a nasty possibility.

The great influenza epidemic of 1918–1919

The outbreak started in the United States and spread to France in April 1918 with arriving American troops. From there, it spread to Spain, where it first caused public alarm (hence the name Spanish flu). It behaved strangely, in that half the victims killed were in the 20–40 age group. The greatest death toll, perhaps 20 million, occurred in India. In a few small communities where everyone fell ill simultaneously, leaving no one to attend the sick, nearly 50% died.

Among people of European descent, about 5 per 1,000 died. The typical death toll among nonwhites was five to ten times higher. In South Africa, many baboons died alongside their human cousins; we cannot be sure, but they probably caught Spanish flu.

It is often suggested that the Spanish flu was somehow caused by the crowded trenches and troopships and was spread by the troop movements of World War I. However, Europe was far more disrupted by World War II, in which both troop and refugee movements occurred on a larger scale than in the earlier war. The crowded air-raid shelters of World War II were ideal breeding grounds for a respiratory disease such as flu. Yet from the beginning of World War II, no major influenza outbreak happened until 1949. A related question is why no lethal flu virus has yet emerged from the massively overcrowded postwar cities of the Third World These questions suggest that Spanish flu was just a chance mutation to a rare virulent form that coincided with the end of World War I. Troop movements probably spread it faster than normal, but if the world had been at peace, the new variant would surely have spread by trade and civilian travel.

As I was writing this chapter, in April 2009, a novel version of swine flu has emerged from Mexico. Despite massive publicity, it has had little real effect, except on the Mexican tourist industry. Although it is aberrant in some ways, so far it is relatively mild. The World Health Organization has declared an official pandemic, based on the worldwide spread of the virus. However, most infected people have recovered without any need for medical care. Mutation to a more virulent form is always possible with flu, but as of now, it seems unlikely that this outbreak will bring about a major disaster.

Disease and the changing climate

The temperature of our planet has fluctuated considerably in the past. We have already mentioned the period of cooling in early medieval times that might have ultimately triggered the Black Death. Ice cores drilled in Greenland indicate that temperatures reached the most recent minimum ("mini Ice Age") about a hundred years ago, when the Thames River that runs through London froze over in winter. Since then, the temperature has been slowly rising. Global warming will have a major impact on infectious disease—mostly for the worse, because disease tends to thrive in hotter moister climates.

One major effect will be to extend the range of mosquitoes and the diseases they carry, especially malaria, yellow fever, and dengue fever. Temperate zones that have been largely malaria-free will likely suffer major intrusions. Deforestation and the creation of large areas of stagnant water by irrigation projects have added to the effects of global warming. Other insect pests and tropical diseases will follow suit.

Global warming coupled with changing rainfall patterns also affects diseases spread by water, including cholera. Outbreaks of disease due to contaminated water in the United States mostly come when rainfall is unusually high. Cholera outbreaks are favored by warmer ocean temperatures and higher rainfall. Extra rainfall increases nutrient run-off from the land into the seas. This drives blooms of marine algae and plankton. This, in turn, allows the cholera bacteria that live inside plankton in coastal waters to proliferate. Cholera in coastal waters is presently moving northward from Peru, where the last major outbreaks occurred.

Floods, which are likely to increase in frequency due to a warmer, wetter climate, tend to aid the spread of disease, especially in poor countries, where hygiene is already dubious. Rodents driven from their homes during floods are apt to spread disease. One example is the outbreak of bubonic plague in Surat, India, in 1994. An earthquake followed by a flood left thousands homeless. Emergency supplies were provided for the human victims. However, hordes of rats were also flooded out and spread plague as they scavenged for food and shelter among the refugees.

Technology-borne diseases

Advances in technology both shut and open doorways for disease. Sewers remove human waste and decrease typhoid and dysentery. Sewers provide highways, or rather subways, for rats to scurry through carrying plague. Virtually every major change in technology has altered the risks of catching some infection. Today is no different. Irrigation projects can cause increased spread of malaria and bilharzia in Africa and Asia.

Changes in food processing have increased food poisoning in the United States. Processing food in ever-larger batches is economically efficient but provides better opportunities for bacteria to spread. Hamburger contaminated with *E. coli* and peanut butter with *Salmonella* have become staple news items in the last few years. Although not novel themselves, many bacteria responsible for food poisoning do carry newly acquired virulence factors. Mad cow disease is one of the few truly new emerging diseases. Its spread was also triggered by changes in animal husbandry.

In contrast, Legionnaire's disease is not new, but its emergence from obscurity did rely on new technology. The bacteria can accumulate in water tanks or cooling towers and spread when humans breathe in the aerosols generated by showers, ventilators, and air-conditioners. Legionnaire's disease was first identified following an outbreak at the American Legion convention in Philadelphia in 1976. Consequently, the bacterium causing it was named *Legionella,* in honor of the American Legion. Since then, sporadic outbreaks of *Legionella* have occurred at hotels and other institutions. Despite the journalistic hype that greeted its emergence on the world stage, *Legionella* is only a minor irritation in global terms. A few hundred cases a year, with a fatality rate of about 10%, occur in the industrial nations. This will probably continue for the foreseeable future.

Emergence of antibiotic resistance

In addition to the threat from truly novel infectious agents, well-established infections can gain new abilities. Since antibiotics were introduced in the 1930s, many bacteria have evolved resistance. Similar problems have been seen with antiviral drugs and with the insecticides used to control insects that carry disease. Before getting too panicky, we should remind ourselves that the great decline in infectious disease happened before antibiotics were discovered. Sanitation and vaccination eliminated most of the dangerous infections from industrial nations.

Antibiotic resistance can appear as a result of a novel mutation or can be transferred from one bacterium to another. Even in the early days of antibiotic use, sporadic cases of resistance arose. Most of these were due to mutations in the bacteria that were being treated, and relatively few of these resistant strains spread. In the absence of the antibiotic, most resistance mutations are harmful to the bacteria. For example, spontaneous mutants of bacteria resistant to streptomycin have defects in protein synthesis. Mutants resistant to kanamycin or neomycin cannot respire properly.

The situation is reminiscent of human mutations that give resistance to malaria but cause sickle-cell anemia or give resistance to typhoid but cause cystic fibrosis. In the absence of the threat (antibiotics for bacteria, malaria for people), the resistant mutants fade away.

The bigger threat comes from transmissible antibiotic resistance. Plasmids are circular segments of extra genetic information that many bacteria carry. Some plasmids move from one bacterial strain to another and carry genes that are "optional extras"—handy under some conditions, but useless or a burden under others. Plasmids can confer the ability to grow on rare and unusual nutrients. They can also carry genes that protect bacteria against antibiotics or toxic metals, both due to human activity. The antibiotic resistance genes carried on plasmids rarely interfere with normal bacterial growth. Instead of risking alterations in vital bacterial genes, plasmids bring in extra genes from outside. These often destroy the antibiotic with no detrimental side effects on the bacteria. Consequently, even in the absence of antibiotics, the antibiotic-resistance plasmids are lost only very slowly. A single plasmid can carry resistance to several antibiotics. Alternatively, a single bacterium can contain several plasmids, each conferring resistance to a single antibiotic. Either way, the result is multiple-antibiotic resistance that can be passed from bacterium to bacterium.

The emergence of antibiotic resistance was inevitable. When living creatures are killed in large numbers, a few resistant individuals usually survive to breed. Nonetheless, the rapid spread of antibiotic resistance has been helped by human greed and stupidity. Farmers often include antibiotics in animal feed. This keeps infection down and supposedly results in more meat per dollar. It also encourages the spread of resistance plasmids that can later transfer to bacteria that infect humans. Many European nations were smart enough to realize that the costs of extra hospital care vastly outweighed the few pennies saved by cheaper bacon and have greatly restricted this practice.

Today agriculture consumes about two-thirds of antibiotics, and only about one-third is used medically. Third World nations are becoming major contributors to this problem, as the increasing demand for meat has led to widespread abuse of antibiotics. While the industrial nations start to clean up their act, many poorer nations are using more antibiotics to increase meat and chicken yields. The only factor in choosing which antibiotics to put in animal feed in poorer countries is the price. This undermines the growing tendency in advanced nations to reserve certain antibiotics for human use.

Another problem is overprescription. Although antibiotics kill only bacteria, doctors often prescribe them for virus infections, such as colds and flu. This abuse is vastly more common in the United States. Partly, Americans want to get something to "cure" them. Explaining that antibiotics don't cure viruses to people with the lowest education standard of any industrial nation is just too much effort. Doctors, in turn, are frightened that if they are honest, they might lose their patients to another, more obliging physician. (In reality, a recent survey by the CDC showed that few patients actually wander from doctor to doctor.) Doctors are also frightened of being sued for failing to provide "appropriate treatment." In England, children with ear or throat infections are given antibiotics only in rare cases when the infection continues. In the United States, "Shoot first, ask questions later" is the motto not just of yesterday's cowboy, but also of today's yuppie parent.

One way to combat resistance is to replace old antibiotics with newly invented ones. Soon after they were first discovered, there was a big rush to discover new antibiotics or modify old ones chemically, yielding new variants. When most known bacterial diseases had cures, complacency set in. Recently, drug resistance has hit the headlines and research has picked up again. Although some new antibiotics are now in the pipeline, it takes several years to get a new drug from laboratory to hospital. As new antibiotics are deployed, resistance will inevitably appear. We can look forward to a permanent cold war between bacteria and pharmaceutical companies.

Where do the resistance genes on plasmids come from? They are gifts from Mother Nature, like most antibiotics. Long before humans isolated penicillin from the mold *Penicillium*, or streptomycin from the bacterium *Streptomyces*, these antibiotics were deployed to wage biological warfare in the soil. Bacteria and molds have been slugging it out for eons before humans joined in the fray. Not only did microorganisms develop antibiotics to kill each other, but they developed resistance mechanisms to counter each other's attacks. Some bacterial cultures stored before penicillin was discovered already had resistance genes. Thus, resistance to most antibiotics probably predates their use by humans. Increased use has led to the spread of these resistance genes.

Disease and the food supply

We have focused on human disease, but remember that livestock and crop plants suffer from infections, too. Modern farmers tend to rely heavily on a few main crops, with little crop rotation. Large areas of a single crop provide the same opportunities for plant diseases that overcrowded cities provide for human infections. The warmer, wetter weather that is becoming more prevalent favors fungal infections that attack plants. For example, wheat scab outbreaks in the United States and Canada caused massive losses in the 1990s.

Decreased surpluses in the major grain exporters undermine the safety net for overpopulated third world nations. If major drought in tropical areas such as Africa or India coincides with major crop losses in the grain exporters, the result could be widespread famine. In 2006–2007, world grain reserves fell to 57 days of consumption, the lowest since 1972.

Perhaps the most serious current threat to our food supply is the wheat rust fungus (*Puccinia graminis*). A new and highly virulent strain emerged from Uganda in 1999 and was, therefore, named Ug99. It is presently in Africa and parts of Asia. Because the spores are airborne, this fungus will inevitably spread worldwide. Breeding resistant wheat varieties is in progress but takes several years.

Overpopulation and microbial evolution

Overpopulation does not merely threaten starvation; it sets the scene for the evolution of new infectious diseases. The more people there are—and the more crowded, unhygienic, and malnourished they are—the greater the opportunity for some new and virulent plague to emerge. So far, we have kept ahead.

A related issue is the growing number of humans with deficient immune systems. Some people are immunocompromised due to drugs used to suppress rejection of organ transplants or drugs used in cancer therapy, but the vast majority are AIDS victims who are infested with a growing variety of opportunistic diseases. Some of these diseases rarely infect healthy people, but others, such as tuberculosis, sometimes infect the healthy. AIDS patients have become evolutionary staging areas where previously harmless microorganisms can adapt to growth in humans without being promptly eradicated by the immune system. As drugs keep AIDS patients alive longer, opportunistic infections get more time to grow and evolve. Long-term antibiotic treatment provides ideal conditions for antibiotic resistance to arise and perhaps spread to other bacteria that are dangerous to people with healthy immune systems.

Predicting the future

> *"Clearly, the future is still to come."*
> —Peter Brooke, member of U.K. Parliament, 1986

Clever prophets take care to make their pronouncements ambiguous. That way, they can claim to be right about whatever happens. Moreover, it doesn't take much insight to realize that wars, earthquakes, famine, and pestilence will make continuing appearances on the world stage.

In his futuristic work *The Shape of Things to Come,* published in 1933, H. G. Wells relies on a novel plague to eliminate half the population of Earth in 1955–1956 and usher in a new era. Although this epidemic was largely modeled on the Black Death, Wells had his "maculated fever" waft around the world on the wind instead of spread by fleas. His fictitious disease emerged from captive baboons in the London Zoological Gardens. *The Shape of Things to Come* was written as a prediction of the future in an age when most scientists foresaw only the eventual eradication of infectious disease, not its resurgence.

So what should we predict? First, let's consider the global situation. The British Empire was the last great civilization.

Improved hygiene, originating from the industrialized West, led to worldwide decreased infant mortality. That, in turn, created a population boom that undermined the profitability of the European colonial empires. Despite poor hygiene and rampant disease relative to the industrial nations, the birth rate still outstrips infant mortality in Third World countries. The ongoing population explosion is the single most important biological trend in today's world.

Denser populations, coupled with poverty, are promoting the spread of disease. Although tuberculosis is in the lead right now, most of those infected do not fall ill. As the remaining sensitive humans are weeded out, the incidence of TB in the Third World will begin to decline naturally, just as it did in Europe a century ago.

In the advanced nations, AIDS will affect homosexuals and intravenous drug users but have marginal impact on the mainstream. Its major effect, especially in the United States, will be to increase the cost of health care in the inner cities. This will help enlarge the growing gap between rich and poor. In Africa and, to a lesser extent, other third world regions, AIDS will thin out the promiscuous and malnourished, and favor the spread of religious puritanism, particularly, Islamic sects.

Still more serious, in my opinion, are malaria and other insect-borne infections that are spreading in the tropics. Rising world temperatures promote the spread of insects that transmit many tropical or subtropical diseases. Human construction and irrigation projects are helping, as is the steady increase in insecticide resistance among the insect carriers. An ugly long-term threat is the possible adaptation of tropical viruses to be carried by insects that survive in colder climates.

Future emerging diseases

The growing Third World cities are the true danger zones for emerging disease. The threat is not so much that Ebola or Lassa might break loose in a crowded slum. More dangerous is the prospect that some disease that already has the capacity to spread effectively might increase in virulence while circulating among the tightly packed masses. A rogue variant of flu or measles that killed a higher proportion of its victims could easily sweep through a crowded Third World city. The denser such populations grow, the greater is the likelihood of such a mutant emerging and spreading.

Such a virus could spread across the world by air travel. As urban decay continues, the cities of industrial nations are gradually becoming more susceptible to such infection. One paradoxical effect of advancing technology is on air pollution. Fumes from automobiles and oil refineries kill most airborne microbes. Clean, pure air allows them to live. Reducing air pollution makes the transmission of airborne infections much easier. Centralized air-conditioning recirculates air, along with any germs it carries, among all the rooms within a building—or an airport complex.

Gloom and doom or a happy ending?

> *"This is the way the world ends*
> *Not with a bang but a whimper."*
> —T. S. Eliot

Until recently, most essays on infectious disease ended on a triumphant note. Human technology has taken care of the problem. Eat, drink, and be merry (at least, until you die of cancer or heart disease)! More recently, the emergence of novel infections, coupled with the problem of increasing antibiotic resistance, has heralded a move to gloom and doom. Perhaps not the next outbreak, nor even the one after that, but soon a plague will emerge that we cannot control. Civilization will collapse, and even if we survive, we will revert to savagery.

Gloom-and-doomers generally opt for a single highly virulent plague that creates unmitigated disaster. However, previous plagues rarely destroyed society as a whole. Instead, they transformed it. Even the medieval Black Death is a case in point. It fits rather well with Nietzsche's maxim: "If it doesn't kill [all of] you, it will make you [society] stronger." Western society emerged improved and less restrictive.

Nonetheless, the Black Death was a terrible disaster, and we certainly do not wish to suffer a parallel experience in the mere hope of future improvement. Thankfully, although providing sufficient resources rapidly is a major problem, the advanced nations have the capacity to keep most foreseeable individual epidemics under reasonable control.

However, as global crowding and travel continue to increase, there will be steadily more novel infections. One can envisage an increasing cumulative disease burden, as opposed to a single devastating plague. In particular, we are swimming in a sea of viruses that constantly mutate. As our populations grow ever denser, we are favoring the emergence of variants of infectious agents with increased virulence.

At the same time, modern technology is spreading from the West to the rest of the world, especially Asia, and is also constantly improving. Essentially, we have become embroiled in a high-tech arms race with the rapidly mutating viruses and, to a lesser extent, with the bacteria, which change more slowly. Although we have suffered some recent setbacks, we are still winning. In most regions of the world, life expectancy and standards of living are increasing, albeit more slowly than in the twentieth century.

The two most populous nations, China and India, both have rapidly developing biotech industries. Indeed, artemisinin, the drug now most favored for treating malaria in the Third World, came from China. Although drug discovery, especially of novel antibiotics, has slowed in the West, I suspect that the emerging high-tech nations will pick up the slack rather soon.

Novel infections will continue to emerge and test our medical technology and health care systems. If we can plot a common-sense course between getting too smug and overreacting to every minor outbreak, I think our chances are pretty good.

Further reading

Fascinating classics written long ago that are still good reading:

Defoe, Daniel. *Journal of the Plague Year.* New York: New American Library, 1960. (Original edition 1723.)

Although a work of fiction, the author lived in times when the bubonic plague was still around.

Nightingale, Florence. *Notes on Nursing: What It Is and What It Is Not.* New York: Dover Publications, 1969. (Original edition 1859.)

For a nice little old lady, Florence Nightingale was amazingly blunt and opinionated. She made generals tremble in their shoes. She would have made Hillary Clinton wilt!

Most important modern works:

Ewald, Paul W. *Evolution of Infectious Disease*. Oxford:
Oxford University Press, 1994.

    Seminal work on the evolution of infectious disease from
the modern genetic and evolutionary viewpoint. Rather
academic.

Herlihy, David. *The Black Death and the Transformation of
the West*. Cambridge, MA: Harvard University Press, 1997.

    Expounds the idea that the Black Death was responsible
for the emergence of Western democracy.

McNeill, W. H. *Plagues and Peoples*. Garden City, NY:
Anchor Press, 1976.

    The most important single source that summarizes and
explains the idea that epidemics affected human history.

Zinsser, Hans. *Rats, Lice & History*. Boston: Little, Brown
and Company, 1934. (Reprinted quite frequently.)

    Classic on typhus fever and history from the viewpoint of
a microbiologist.

Narrow in focus, yet fascinating:

Cantor, Norman F. *In the Wake of the Plague*. New York:
Free Press, 2001.

    How the Black Death remodeled European society.

Cockburn, Aidan, and Eve Cockburn. *Mummies, Disease
and Ancient Cultures*. Cambridge, U.K.: Cambridge
University Press, 1980.

Grmek, Mirko D. *Diseases in the Ancient Greek World*.
Baltimore: Johns Hopkins University Press, 1989.

A selection of other interesting books:

Cartwright, Frederick F., and Michael D. Biddiss. *Disease and History.* New York: Dorset Press, 1972.

Crawford, Dorothy H. *Deadly Companions.* Oxford: Oxford University Press, 2007.

Diamond, Jared. *Guns, Germs and Steel.* New York: W. W. Norton, 1998.

Garrett, Laurie. *The Coming Plague.* New York: Penguin Books, 1995.

Oldstone, Michael B. A. *Viruses, Plagues, and History.* New York: Oxford University Press, 1998.

Preston, Richard. *The Hot Zone.* New York: Random House, 1994.

Wills, Christopher. *Yellow Fever, Black Goddess: The Coevolution of People and Plagues.* Reading, MA: Addison-Wesley, 1996. (First published in the United Kingdom by HarperCollins as *Plagues: Their Origins, History and Future.*)

Websites that deal with epidemics and infections:

http://www.cdc.gov/
Centers for Disease Control

http://www.who.int/csr/don/en/
World Health Organization disease outbreak news

http://www.fda.gov/Food/FoodSafety/FoodborneIllness/
FoodborneIllnessFoodbornePathogensNaturalToxins/
BadBugBook/default.htm
FDA site about foodborne disease

http://fas.org/irp/threat/cbw/
Federation of American Scientists on biological and
chemical weapons

http://www.ifrc.org/
Red Cross and Red Crescent on disasters, including
epidemics

http://www.mic.stacken.kth.se/Diseases/
Archive on disease from the Karolinska Institute

Index

Plasmodium falciparum, 15, 45
pneumonic plague, mortality
rate, 187
poisoning. *See also* food
poisoning
accidental versus intentional, 34
ergot poisoning, 106-109
poliomyelitis, 225-226
resistance to, 225
transmission method, 39
political response to mad cow
disease, 101-102
polytheism
in Middle Ages, 188
monotheism versus, 179-181
Pommery, Madame, 137
population collapse, periods of,
231-232
population density. *See also*
ancient civilizations; cities
accuracy of estimates, 220
after bubonic plague, 218-219
of ancient Athens, 28
during Middle Ages, 217-218
effects of
*on disease transmission,
56-59*
on disease virulence, 41-44
efficiency of transmission
methods and, 242
European empires versus Third
World nations, 226
future predictions, 253
in imperial expansion, 120-122
indigenous Americans, 197
opportunistic diseases and,
251-252
replenishing in cities, 213-214
required for measles, 25
in Roman Empire, 82
susceptibility to disease and, 8
population statistics, 231

positive aspects of epidemics,
6-8, 53, 56
potato blight, 109-110
pre-Columbian Americans. *See*
indigenous Americans
predicting the future, 252-256
prion disease, 98-99
mad cow disease
in humans, 102-103
origin of, 99-101
*political response to,
101-102*
scrapie, 99
prion protein, resistance to mad
cow disease, 55
Procopius, 89, 91
promiscuity
puritanism versus, 141-143
rates of, propaganda versus,
144-145
propaganda, actual promiscuity
rates versus, 144-145
prostitution
sacred prostituion, 160-161
tolerance for, 154
in Victorian-era England, 142
protection from evil spirits,
178-179
protozoa, mutation rate, 61
protozoan diseases, rate of
evolution, 24
Prusiner, Stanley, 98
psychological effects of biological
warfare, 131
public health
AIDS and, 156, 158
breakdowns in, 236
Puccinia graminis, 251
punishment, disease as, 179-181
puritanism, promiscuity versus,
141-143
Puritans, 202-203

X–Y–Z

FINANCIAL TIMES

In an increasingly competitive world, it is quality
of thinking that gives an edge—an idea that opens new
doors, a technique that solves a problem, or an insight
that simply helps make sense of it all.

We work with leading authors in the various arenas
of business and finance to bring cutting-edge thinking
and best-learning practices to a global market.

It is our goal to create world-class print publications
and electronic products that give readers
knowledge and understanding that can then be
applied, whether studying or at work.

To find out more about our business
products, you can visit us at www.ftpress.com.